Designer Notes for Microwave Antennas

For a complete listing of the *Artech House Microwave Library*,
turn to the back of this book . . .

Designer Notes for Microwave Antennas

Richard C. Johnson

Artech House
Boston • London

Library of Congress Cataloging-in-Publication Data

Johnson, Richard C. (Richard Clayton), 1930-
Designer notes for microwave antennas / Richard C. Johnson.
p. cm.
Includes bibliographical references and index.
ISBN 0-89006-521-7
1. Microwave antennas. I. Title.
TK7871.6.J64 1991 91-2134
621.382'4--dc20 CIP

British Library Cataloguing in Publication Data

Johnson, Richard C.
Designer notes for microwave antennas.
1. Microwave equipment. Antennas (Radio) - Design
I. Title.
621.38133

ISBN 0-89006-521-7

685 Canton Street
Norwood, MA 02062

International Standard Book Number: 0-89006-521-7
Library of Congress Catalog Card Number: 91-2134

10 9 8 7 6 5 4 3 2 1

CONTENTS

PREFACE

My career with the Georgia Tech Research Institute (then known as the Engineering Experiment Station) started back in the 1950s. A small group of us who were working on radar antennas and microwave components decided to keep design notebooks to assist ourselves in future design tasks. Each time one of us developed or discovered useful design information, we documented the information and made a copy for each member of the group.

As the group grew larger, our internal communication decreased (as is usually the case), and soon the information exchanges were forgotten. However, I found my design notebook to be so useful that I continued to collect and add new data when available. On many occasions, I took along excerpts from my notebook on trips when I needed quick access to design information.

Most microwave antenna design problems involve the use of known techniques—even for engineers working in research and development organizations. The first "task" is to locate appropriate design information for solving the problem.

On such occasions, I usually reach for my notebook first. Because I find the notes to be so useful, I have decided to "smooth out" and supplement them for presentation in this book so that others can have easy access to the collection of information. If this book enables you to save a few hours during a routine design procedure, then the book will have served its purpose.

Chapters 1 and 2 present some basic concepts, definitions, rules of thumb, and far-field characteristics for some theoretical aperture distributions. Chapter 3 provides equations for many characteristics of waves in several types of transmission lines. Chapters 4 through 7 are devoted to design data for popular microwave antennas and radomes. Impedance and matching techniques are presented in Chapter 8, and some topics related to antenna measurements are discussed in Chapter 9. Some techniques for extracting information from voltage standing wave ratios and radiation patterns are given in Chapter 10. A broad selection of special topics is contained in the Appendix. The book is a collection of information that has interested me during my career; because of the way in which the data were collected, subjects are not discussed to a uniform depth.

I have attempted to present design data succinctly with enough narrative to describe how to use the information and with appropriate references for locating additional sources, if needed. Design curves are presented in large graphs with cross-hatching or tick-marks to assist in extracting specific data. Each chapter contains a list that defines symbols used in the chapter; thus, the reader does not have to flip back and forth through the book searching for a definition.

I want to thank Rickey Cotton for reviewing the first draft of this book and Arch Corriher for reading the last draft and for making many suggestions during the project; both are engineers with the Georgia Tech Research Institute.

I also thank the many publishers who have granted permission to use material from their publications. I have tried to credit all sources of information with references; any omissions are due to oversight rather than intent.

RICHARD C. JOHNSON
DALLAS, TEXAS
NOVEMBER 1990

Chapter 1
INTRODUCTION—CONCEPTS AND DEFINITIONS

Consider an antenna located at the origin of a spherical coordinate system as illustrated in Figure 1.1. Suppose that we are making observations on a spherical shell having a very large radius r. Assume that the antenna is transmitting, and let

- P_0 = power accepted from transmission line, W
- P_r = power radiated by antenna, W
- η = radiation efficiency, unitless

The above quantities are related as follows:

$$\eta = \frac{P_r}{P_0} \tag{1.1}$$

Let

- $\Phi(\theta,\phi)$ = radiation intensity, W/sr

Note that because r was assumed to be very large, Φ is independent of r. This independence of r is a characteristic of the far-field region. The total power radiated from the antenna is

$$P_r = \int_0^{2\pi} \int_0^{\pi} \Phi(\theta,\phi) \sin\theta \, d\theta \, d\phi \tag{1.2}$$

and the average radiation intensity is

$$\Phi_{avg} = \frac{P_r}{4\pi} \tag{1.3}$$

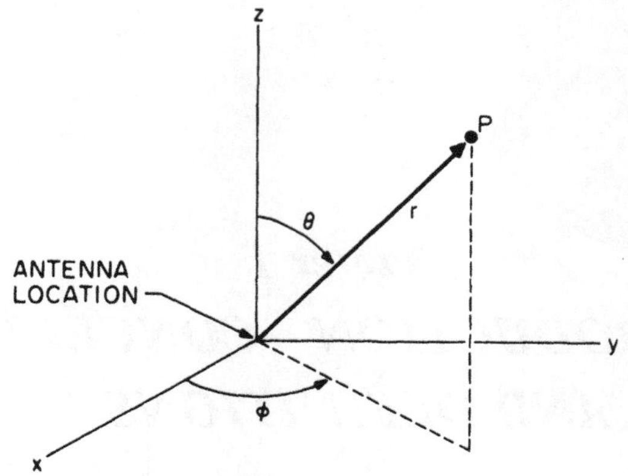

Figure 1.1 An antenna in a spherical coordinate system. (From Johnson [1], with permission of McGraw-Hill, Inc.)

Let

- $D(\theta,\phi)$ = directivity, unitless

Directivity is a measure of the ability of an antenna to concentrate radiated power in a particular direction, and it is related to the radiation intensity as follows:

$$D(\theta,\phi) = \frac{\Phi(\theta,\phi)}{\Phi_{avg}} = \frac{\Phi(\theta,\phi)}{P_r/4\pi} \tag{1.4}$$

The directivity of an antenna is the ratio of the achieved radiation intensity in a particular direction to that of an isotropic antenna. In practice, we are usually interested primarily in the peak directivity of the main lobe. Thus, if we say that an antenna has a directivity of 100 (or 20 dB), 100 is assumed to be the peak directivity of the main lobe.

Let

- $G(\theta,\phi)$ = gain, unitless

The gain of an antenna is related to the directivity and power radiation intensity as follows:

$$\begin{aligned} G(\theta,\phi) &= \eta D(\theta,\phi) \\ &= \frac{\eta\Phi(\theta,\phi)}{P_r/4\pi} \end{aligned} \tag{1.5}$$

and from eq. (1.1),

$$G(\theta,\phi) = \frac{\Phi(\theta,\phi)}{P_0/4\pi} \tag{1.6}$$

Thus, the gain is a measure of the ability to concentrate the power accepted by the antenna in a particular direction. Note that if we have a lossless antenna (i.e., η = 1), the directivity and the gain are identical.

Let

- $P(\theta,\phi)$ = power density, W/m^2

The power density is related to the radiation intensity as follows:

$$P(\theta,\phi) = \frac{\Phi(\theta,\phi) \sin\theta\Delta\theta\Delta\phi}{(r\Delta\theta)(r \sin\theta\Delta\phi)}$$

or

$$P(\theta,\phi) = \frac{\Phi(\theta,\phi)}{r^2} \tag{1.7}$$

Substituting eq. (1.6) into (1.7) yields

$$P(\theta,\phi) = G(\theta,\phi)\frac{P_0}{4\pi r^2} \tag{1.8}$$

The factor $P_0/4\pi r^2$ represents the power density that would result if the power accepted by the antenna were radiated by a lossless isotropic antenna.

- $A_e(\theta,\phi)$ = effective area, m^2

Visualizing the concept of effective area is easier if we consider a receiving antenna; it is a measure of the effective absorption area presented by an antenna to an incident plane wave. The effective area is related [2] to gain and wavelength as follows:

$$A_e(\theta,\phi) = \frac{\lambda^2}{4\pi} G(\theta,\phi) \tag{1.9}$$

Many high-gain antennas such as horns, reflectors, and lenses are said to be *aperture-type antennas*. The aperture is usually considered to be that portion of a plane surface near the antenna, perpendicular to the direction of maximum radiation, through which most of the radiation flows. Let

- η_a = antenna efficiency of aperture-type antenna, unitless
- A = physical area of antenna's aperture, m^2

Then,

$$\eta_a = \frac{A_e}{A} \tag{1.10}$$

The term η_a sometimes is called *aperture efficiency.*

When dealing with aperture antennas, we see from eqs. (1.9) and (1.10) that

$$G = \eta_a \frac{4\pi}{\lambda^2} A \tag{1.11}$$

The term η_a actually is the product of several factors, such as

$$\eta_a = \eta\eta_i\eta_1\eta_2\eta_3 \ldots \tag{1.12}$$

The term η is radiation efficiency as defined in eq. (1.1). The term η_i is aperture illumination efficiency, which is a measure of how well the aperture is utilized for collimating the radiated energy; it is the ratio of the directivity that is obtained to the standard directivity. The standard directivity is obtained when the aperture is excited with a uniform, equiphase distribution. (Such a distribution yields the highest directivity of all equiphase excitations.) For planar apertures in which $A \gg \lambda^2$, the standard directivity is $4\pi A/\lambda^2$, with radiation confined to a half space.

The other factors, $\eta_1,\eta_2,\eta_3, \ldots$, include all other effects that reduce the gain of the antenna. Examples are spillover losses in reflector or lens antennas, phase-error losses due to surface errors on reflectors or random phase errors in phased-array elements, aperture blockage, depolarization losses, *et cetera.*

Polarization (see Reference [3] for a more detailed discussion) is a property of a single-frequency electromagnetic wave; it describes the shape and orientation of the locus of the extremity of the field vectors as a function of time. In antenna engineering, we are primarily interested in the polarization properties of plane waves or of waves that can be considered to be planar over the local region of observation. For plane waves, we need only specify the polarization properties of the electric field vector because the magnetic field vector is simply related to the electric field vector.

The plane containing the electric and magnetic fields is called the *plane of polarization,* and it is perpendicular to the direction of propagation. In the general case, the tip of the electric field vector moves along an elliptical path in the plane of polarization. The polarization of the wave is specified by the shape and orientation of the ellipse and the direction in which the electric field vector traverses the ellipse.

The shape of the ellipse is specified by its *axial ratio*—the ratio of the major axis to the minor axis. The orientation is specified by the *tilt angle*—the angle between the major axis and a reference direction when viewed in the direction of propagation. The direction in which the electric field vector traverses the ellipse is the *sense of polarization*—right-handed (clockwise) or left-handed (counterclockwise) when viewed in the direction of propagation.

The polarization of an antenna in a specific direction is defined to be the polarization of the far-field wave radiated in that direction from the antenna. Usually, the polarization of an antenna is relatively constant throughout the main lobe, but it varies much in the minor lobes.

For convenience, we define a spherical coordinate system associated with an antenna as illustrated in Figure 1.2. The polarization ellipse for the direction (θ,ϕ) is shown inscribed on a spherical shell surrounding the antenna. A common practice is to choose $\mathbf{u}_\theta$ (the unit vector in the θ direction) as the reference direction. The tilt angle then is measured from $\mathbf{u}_\theta$ toward $\mathbf{u}_\phi$. The sense of polarization is right-handed if the electric field vector traverses the ellipse from $\mathbf{u}_\theta$ toward $\mathbf{u}_\phi$ as viewed in the direction of propagation, and left-handed if the reverse is true.

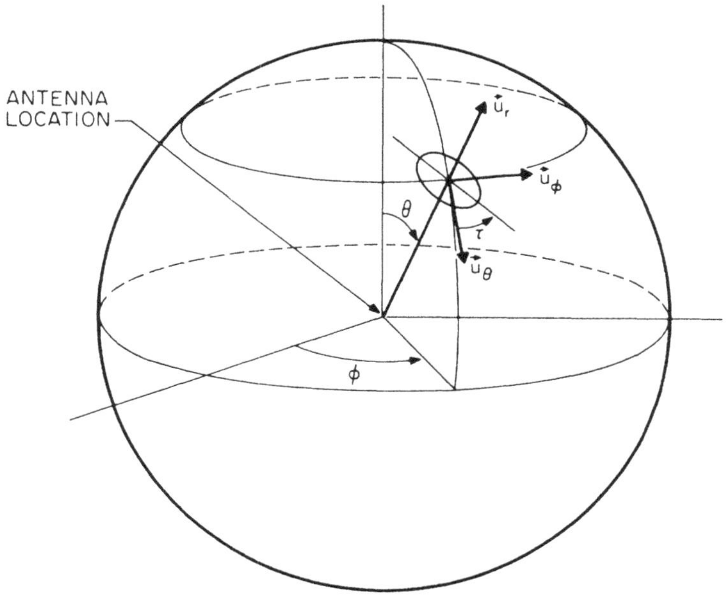

Figure 1.2 Polarization ellipse in relation to antenna coordinate system. (From Johnson [1], with permission of McGraw-Hill, Inc.)

In many practical situations, such as antenna measurements, the establishment of a local coordinate system is convenient. Usually, the u_3 axis is the direction of propagation, the u_1 axis is horizontal, and the u_2 axis is orthogonal to the other two so that the unit vectors are related by $\mathbf{u}_1 \times \mathbf{u}_2 = \mathbf{u}_3$. The tilt angle is measured from $\mathbf{u}_1$.

When an antenna receives a wave from a particular direction, the response will be greatest if the polarization of the incident wave has the same axial ratio, the same sense of polarization, and the same spatial orientation as the polarization of the antenna in that direction. The situation is depicted in Figure 1.3 where E_t represents a transmitted wave (antenna polarization) and E_m represents a matched incident wave. Note that the senses of polarization for E_t and E_m are the same when viewed in their local coordinate systems. Also, note that the tilt angles are different because the directions of propagation are opposite. As depicted in Figure 1.3, τ_t is the tilt angle of the transmitted wave and τ_m is the tilt angle of the polarization-matched received wave; they are related by

$$\tau_m = \pi - \tau_t \tag{1.13}$$

The polarization of the matched incident wave, as described above, is called the *receiving polarization* of the antenna.

When the polarization of the incident wave is different from the receiving polarization of the antenna, then a loss due to polarization mismatch occurs. Let

- η_p = polarization efficiency, unitless

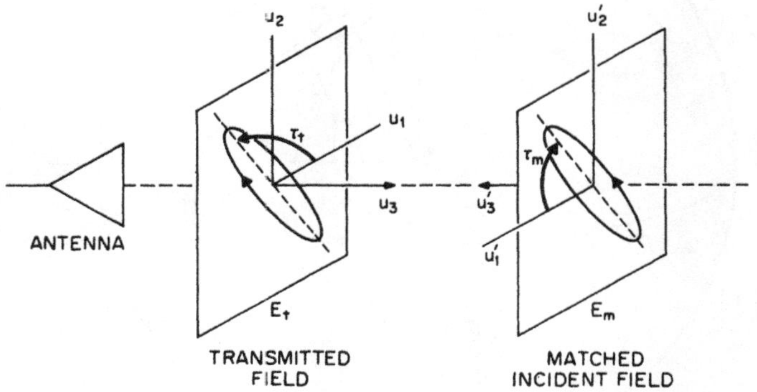

Figure 1.3 Relation between polarization properties of an antenna when transmitting and receiving. (Reproduced from [3], ©1978 by the IEEE, with the permission of the IEEE Standards Department.)

The polarization efficiency is the ratio of the power actually received by the antenna to the power that would be received if the polarization of the incident wave were matched to the receiving polarization of the antenna.

The Poincaré sphere, as shown in Figure 1.4, is a convenient representation of polarization states. Each possible polarization state is represented by a unique point on the unit sphere. Latitude represents axial ratio, with the poles being circular polarizations; the upper hemisphere is for left-handed sense, and the lower hemisphere is for right-handed sense. Longitude represents tilt angles from 0 to π. An interesting feature of the Poincaré sphere is that diametrically opposite points represent orthogonal polarizations.

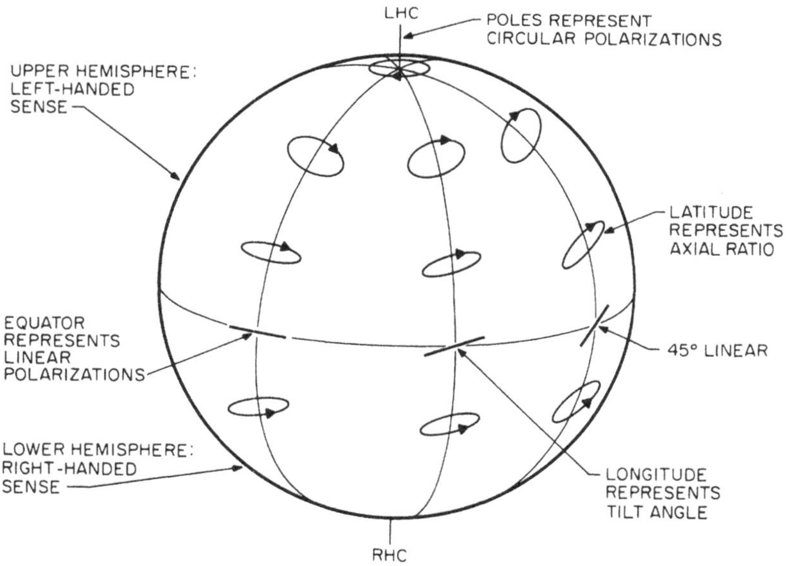

Figure 1.4 Polarization states on the Poincaré sphere. (Reproduced from [3], ©1978 by the IEEE, with the permission of the IEEE Standards Department.)

The Poincaré sphere is also convenient for representing polarization efficiency. In Figure 1.5, W represents the polarization of an incident wave, and A_r represents the receiving polarization of the antenna. If the angular distance between the two points is 2ξ, then the polarization efficiency is

$$\eta_p = \cos^2\xi \tag{1.14}$$

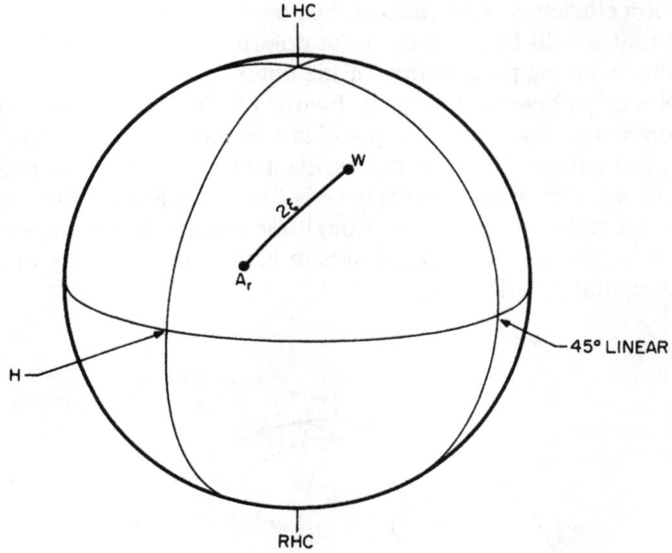

Figure 1.5 Receiving polarization of an antenna A_r and polarization of an incident wave *W*. (From Johnson [1], with permission of McGraw-Hill, Inc.)

REFERENCES

1. R.C. Johnson, "Introduction to Antennas," Chapter 1 in *Antenna Engineering Handbook,* 2nd Ed. (R.C. Johnson and H. Jasik, eds.), New York: McGraw-Hill, 1984.
2. S. Silver, *Microwave Antenna Theory and Design,* MIT Radiation Laboratory Series, Vol. 12, New York: McGraw-Hill, 1949, Sec. 2.14.
3. *IEEE Standard Test Procedures for Antennas,* IEEE Standard 149-1979, New York: Institute of Electrical and Electronics Engineers, 1979, Sec. 11.

LIST OF SYMBOLS

A	physical area of antenna's aperture, m^2
$A_e(\theta,\phi)$	effective area, m^2
$D(\theta,\phi)$	directivity, unitless
E_m	matched incident field
E_t	transmitted field
$G(\theta,\phi)$	gain, unitless
$P(\theta,\phi)$	power density, W/m^2
P_0	power accepted from transmission line, W

P_r	power radiated by antenna, W
r	spherical coordinate
$\mathbf{u}_\theta, \mathbf{u}_\phi$	unit vectors in θ and ϕ directions, respectively
$\mathbf{u}_1, \mathbf{u}_2, \mathbf{u}_3$	unit vectors in local coordinate system
η	radiation efficiency, unitless
η_a	antenna efficiency of an aperture-type antenna, unitless
η_i	aperture illumination efficiency
η_p	polarization efficiency, unitless
θ	spherical coordinate
τ_m	tilt angle of the polarization-matched received wave
τ_t	tilt angle of the transmitted wave
ϕ	spherical coordinate
$\Phi(\theta,\phi)$	radiation intensity, W/sr

Chapter 2
BASIC DESIGN TECHNIQUES

2.1 ROUGH DESIGNS

In many situations, rough designs for aperture antennas are desirable. These rough designs can be established conveniently with a few rules of thumb. The *first* rule is

$$\theta_3 \approx 70 \frac{\lambda}{D} \tag{2.1}$$

where

θ_3 = 3-dB beamwidth in degrees
λ = wavelength
D = largest aperture dimension in the plane of the beamwidth

Figure 2.1 illustrates solutions of eq. (2.1).

The *second* rule relates gain and beamwidth as

$$G \approx \frac{30{,}000}{\theta_{3a}\theta_{3e}} \tag{2.2}$$

where

G = antenna gain
θ_{3a} = 3-dB beamwidth in degrees in the azimuth plane
θ_{3e} = 3-dB beamwidth in degrees in the elevation plane

Strictly speaking, we have assumed that the principal planes of the beam are in azimuth and elevation; if this is not the case, use the principal-plane beamwidths. Figure 2.2 illustrates solutions of eq. (2.2).

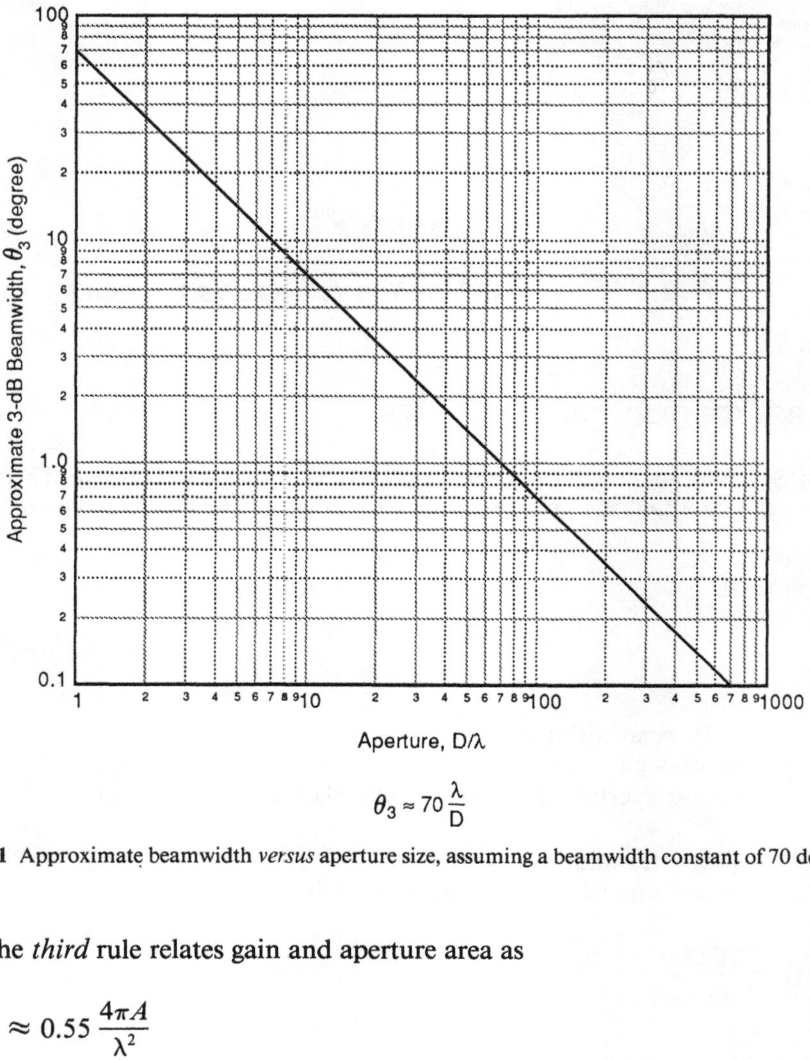

Figure 2.1 Approximate beamwidth *versus* aperture size, assuming a beamwidth constant of 70 degrees.

The *third* rule relates gain and aperture area as

$$G \approx 0.55 \frac{4\pi A}{\lambda^2} \tag{2.3}$$

where A is the aperture area.

The rules assume that the antenna has low losses (high radiation efficiency), hence, gain approximately equals directivity; this is usually true for aperture antennas. If, however, the antenna has significant losses, then gain in the above rules should be replaced by directivity.

To relate the above rules, consider the following example. Suppose that we have a paraboloidal reflector with $D = 1$ m and that we are operating at 10 GHz ($\lambda = 0.03$ m). From eq. (2.1), $\theta_3 \approx 2.1$ degrees; from eq. (2.2), $G \approx 6803$ (or 38.3

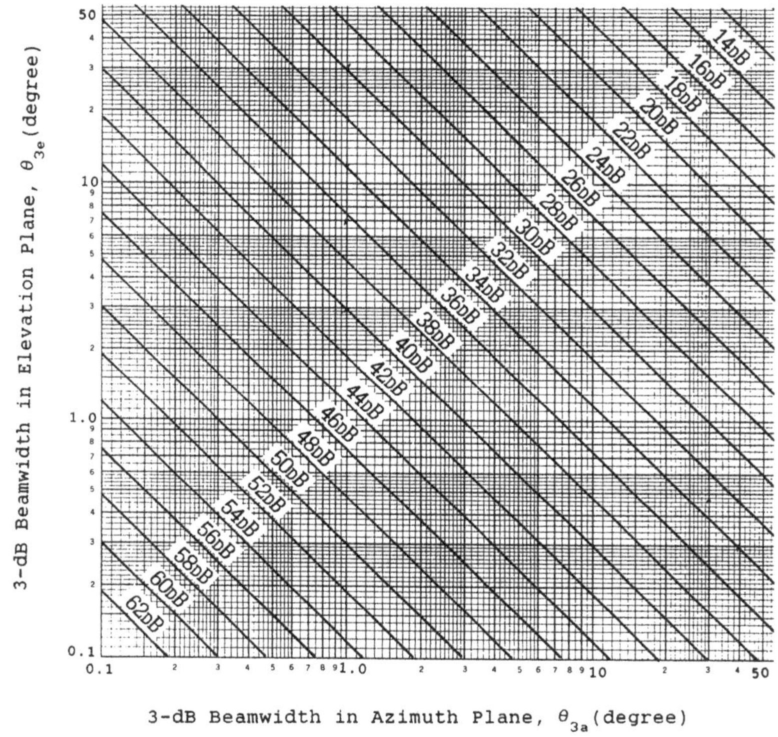

Figure 2.2 Approximate gain *versus* beamwidth, assuming a gain constant of 30,000. The approximation is more accurate for narrow beamwidths.

dBi); and from eq. (2.3), G $\approx$ 6031 (or 37.8 dBi). The 0.5-dB difference in the gain estimates is easily acceptable in rough estimates.

As another example, suppose that we want a circular paraboloidal-reflector antenna operating at 10 GHz and a gain of 40 dBi (or 10,000). From eq. (2.3), $A \approx 1.30\ m^2$, and $D \approx 1.29$ m; from eq. (2.1), $\theta_3 \approx 1.63$ degrees. As an alternative, from eq. (2.2), $\theta_3 \approx 1.73$ degrees, and from eq. (2.1), $D \approx 1.21$ m.

2.2 IMPROVED ESTIMATES

The three equations discussed above are handy for rough estimates, but the means exist to improve the accuracy of these estimates. Equation (2.1) can be written as

$$\theta_3 = k\frac{\lambda}{D} \tag{2.4}$$

where k is the beamwidth constant in degrees. The value of k depends primarily on the aperture illumination function. (For the rough designs above, we set $k = 70$ degrees.) Generally speaking (but not always), illumination functions that yield lower sidelobes result in larger values of k.

Komen [1] reported on the variation of the beamwidth constant for reflector-type antennas. From computed patterns for various edge illuminations, he determined that

$$k = 1.05238I + 55.9486 \tag{2.5}$$

where I is the absolute value of edge illumination (including space attenuation) in decibels and k is in degrees. (In practice, we would normally calculate k to only a few significant figures.) By applying eqs. (2.4) and (2.5) to measured data from several antennas, Komen concluded that the relationship between beamwidth and edge illumination holds regardless of frequency, reflector size, reflector type, or feed type.

The beamwidth constant *versus* edge illumination for reflector-type antennas is illustrated in Figure 2.3. The approximate sidelobe level (SLL) is shown as a solid line for plane-wave feeds and as a dashed line for horn feeds with a parabolic primary pattern. The beamwidth constant and the sidelobe level depend primarily on

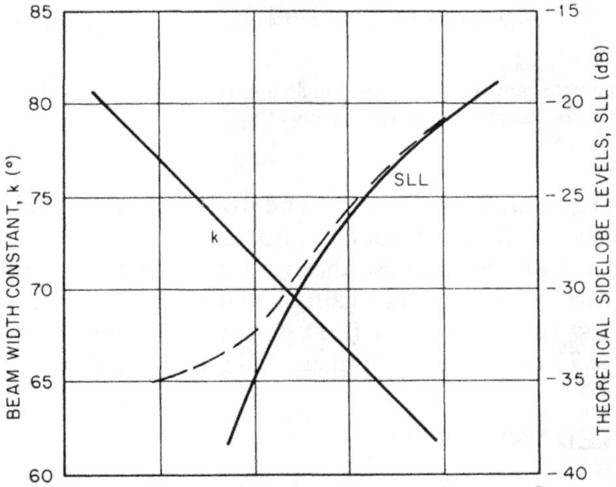

Figure 2.3 Beamwidth constant and approximate first sidelobe level *versus* edge illumination (including space attenuation) for paraboloidal-reflector antennas. The SLL curve is solid for plane-wave feeds and dashed for feeds having parabolic primary patterns (dB *versus* angle).

the edge illumination, but the SLL also depends significantly on the shape of the illumination function.

Similarly, eq. (2.2) can be written:

$$G \approx \frac{K}{\theta_{3a}\theta_{3e}} \tag{2.6}$$

where K is a unitless gain constant. The correct value of K for an actual antenna depends on antenna efficiency; several values are in use. For example, Stutzman and Thiele [2] suggest a value of 26,000, and Stegan [3] suggests a value of 35,000. I usually use 30,000 as in eq. (2.2).

2.3 BEAMWIDTH CONVERSIONS

Beamwidths are usually reported at the 3-dB level for secondary patterns and at the 10-dB level for primary (feed) patterns. The beamwidths at other levels can be esti-

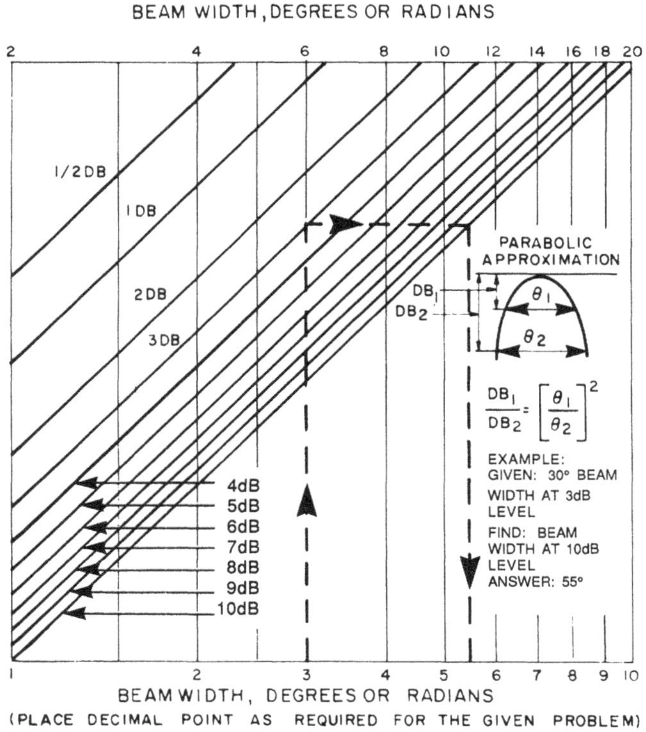

Figure 2.4 Beamwidth conversion chart, assuming parabolic pattern shapes. (From [4].)

mated by assuming that the beamshape is parabolic (when plotted as power in dB *versus* pattern angle) as illustrated in Figure 2.4. As indicated, the beamwidths and power levels are related by

$$\frac{\mathrm{dB}_1}{\mathrm{dB}_2} = \left(\frac{\theta_1}{\theta_2}\right)^2 \tag{2.7}$$

Figure 2.4 can be used to make beamwidth conversions graphically.

2.4 THEORETICAL APERTURE DISTRIBUTIONS

At times, examining the effects on pattern characteristics resulting from various theoretical aperture distributions is instructive. Some classic cases (with uniform phase) have been presented by Silver [5]. Table 2.1 is valid for line-source apertures or for rectangular apertures with separable illumination functions. Table 2.2 is valid for circular apertures with the indicated circularly symmetric illumination functions. These tables demonstrate generally that as the illumination taper increases (reduced edge illumination), the gain decreases, the beamwidth increases, and the sidelobes decrease. Also, note that uniform amplitude illumination results

Table 2.1
Pattern Characteristics Produced by the Indicated Field Distributions for Line-Source Apertures or for Rectangular Apertures with Separable Illumination Functions*

	Gain Factor	*Half-Power Beamwidth*	*Position of First Null*	*First Sidelobe (dB)*
		$f(x) = 1 - (1 - \Delta)x^2$		
$\Delta = 1.0$	1.000	$0.88\lambda/a$	$1.00\lambda/a$	−13.2
0.8	0.994	$0.92\lambda/a$	$1.06\lambda/a$	−15.8
0.5	0.970	$0.97\lambda/a$	$1.14\lambda/a$	−17.1
0.0	0.833	$1.15\lambda/a$	$1.43\lambda/a$	−20.6
		$f(x) = \cos^n(\pi x/2)$		
$n = 0$	1.000	$0.88\lambda/a$	$1.0\lambda/a$	−13.2
1	0.810	$1.20\lambda/a$	$1.5\lambda/a$	−23
2	0.667	$1.45\lambda/a$	$2.0\lambda/a$	−32
3	0.575	$1.66\lambda/a$	$2.5\lambda/a$	−40
4	0.515	$1.93\lambda/a$	$3.0\lambda/a$	−48

*After Silver [5], p. 187.
Notes: $f(x)$ = aperture field distribution, $-1 \leq x \leq 1$, and a = aperture width.

Table 2.2
Pattern Characteristics Produced by a Field Distribution $(1 - r^2)^p$ over a Circular Aperture*

p	*Gain Factor*	*Half-Power Beamwidth*	*Position of First Null*	*First Sidelobe (dB)*
0	1.00	$1.02\lambda/D$	$\sin^{-1} 1.22\lambda/D$	−17.6
1	0.75	$1.27\lambda/D$	$\sin^{-1} 1.63\lambda/D$	−24.6
2	0.56	$1.47\lambda/D$	$\sin^{-1} 2.03\lambda/D$	−30.6
3	0.44	$1.65\lambda/D$	$\sin^{-1} 2.42\lambda/D$	
4	0.36	$1.81\lambda/D$	$\sin^{-1} 2.79\lambda/D$	

*After Silver [5], p. 195.
Notes: r = radial coordinate of aperture, $0 \le r \le 1$, and D = diameter of aperture.

in −13.2-dB sidelobes from a line source and −17.6-dB sidelobes from a circular aperture.

The beamwidth constants and the aperture illumination efficiencies for several continuous line-source distributions are presented [6] in Figures 2.5 and 2.6, respectively. The Chebyschev distribution is not practical because the remote sidelobes do not decay in amplitude, but it is shown for comparison.

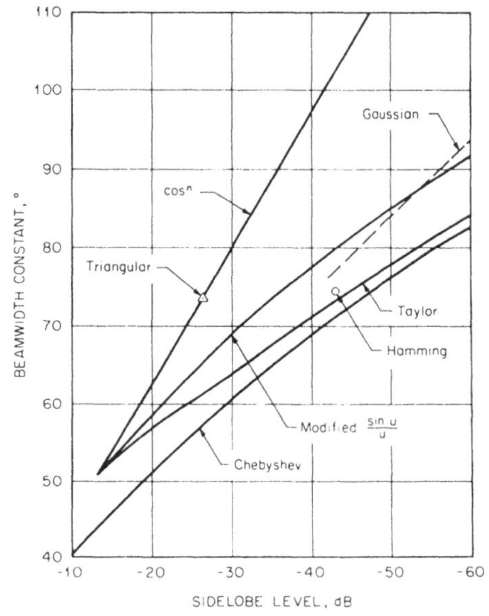

Figure 2.5 Beamwidth constant *versus* sidelobe level for several line-source aperture distributions. (From Bodnar [6], with permission of McGraw-Hill, Inc.)

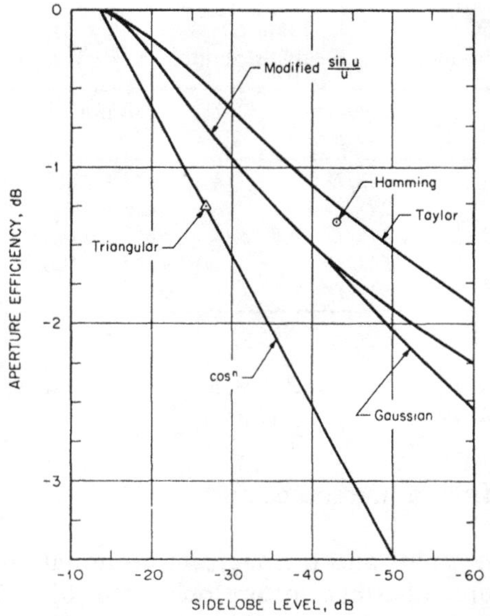

Figure 2.6 Aperture efficiency *versus* sidelobe level for several line-source aperture distributions. (From Bodnar [6], with permission of McGraw-Hill, Inc.)

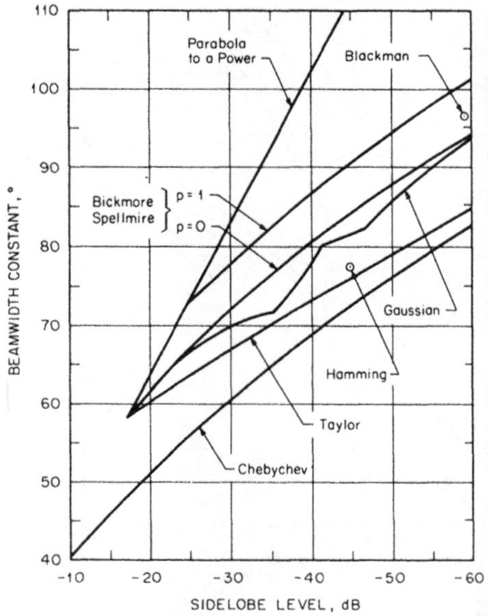

Figure 2.7 Beamwidth constant *versus* sidelobe level for several circular-aperture distributions. (From Ludwig [7].)

The beamwidth constants and the aperture illumination efficiencies for several continuous circular-aperture distributions are presented [7] in Figures 2.7 and 2.8, respectively.

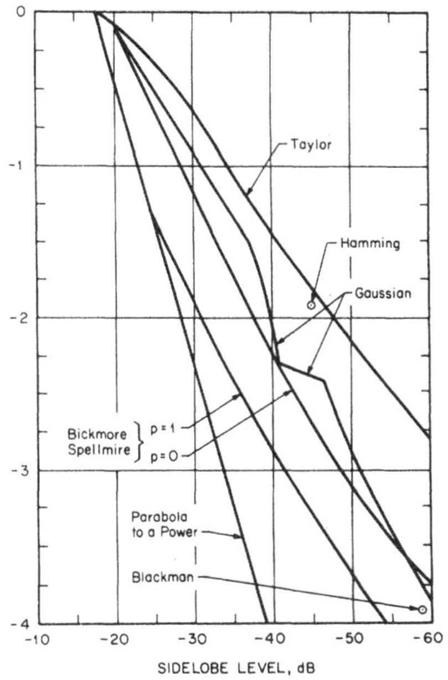

Figure 2.8 Aperture efficiency *versus* sidelobe level for several circular-aperture distributions. (From Ludwig [7].)

REFERENCES

1. M.K. Komen, "Use Simple Equations to Calculate Beamwidth," *Microwaves,* December 1981, pp. 61–63.
2. W.L. Stutzman and G.A. Thiele, *Antenna Theory and Design,* New York: John Wiley and Sons, 1981, p. 397.
3. R.J. Stegan, "The Gain-Beamwidth Product of an Antenna," *IEEE Trans. Antennas and Propagation,* Vol. AP-12, July 1964, pp. 505–506.
4. *ITE Antenna Handbook,* 2nd Ed., Philadelphia: ITE Circuit Breaker Company, *circa* 1965.
5. S. Silver, *Microwave Antenna Theory and Design,* MIT Radiation Laboratory Series, Vol. 12, New York: McGraw-Hill, 1949, pp. 187, 195.
6. D.G. Bodnar, "Materials and Design Data," Chapter 46 in *Antenna Engineering Handbook,* 2nd Ed. (R.C. Johnson and H. Jasik, eds.), New York: McGraw-Hill, 1984.
7. A.C. Ludwig, "Low Sidelobe Aperture Distributions for Blocked and Unblocked Circular Apertures," RM 2367, General Research Corporation, April 1981. Also see *IEEE Trans. Antennas and Propagation,* Vol. AP-30. September 1982, pp. 933–946.

LIST OF SYMBOLS

A	aperture area
a	aperture width
D	largest aperture dimension in the plane of the beamwidth
G	antenna gain
I	absolute value of edge illumination (including space attenuation), dB
K	gain constant
k	beamwidth constant, degrees
SLL	sidelobe level
θ	pattern angle
θ_3	3-dB beamwidth, degrees
θ_{3a}	3-dB beamwidth in azimuth plane, degrees
θ_{3e}	3-dB beamwidth in elevation plane, degrees
λ	wavelength

Chapter 3
TRANSMISSION-LINE CHARACTERISTICS

3.1 WAVES BETWEEN PARALLEL PLANES

Consider a pair of parallel, perfectly conducting planes of infinite extent in the y and z directions with a nonconducting region between the planes as illustrated in Figure 3.1. We assume that propagation is in the z direction and that the variation of all field components in this direction may be expressed in the form

$$e^{-\gamma z}$$

where, in general,

$$\gamma = \alpha + j\beta$$

is a complex propagation constant. (See the list of symbols at the end of this chapter for the definitons of symbols.)

3.1.1 TE and TM Modes

1. Field Equations:

$$\left.\begin{aligned} E_y &= C_1 \sin\left(\frac{m\pi}{a}x\right) e^{j(\omega t - \beta z)} \\ H_x &= \frac{-\beta}{\omega\mu} C_1 \sin\left(\frac{m\pi}{a}x\right) e^{j(\omega t - \beta z)} \\ H_z &= \frac{jm\pi}{\omega\mu a} C_1 \cos\left(\frac{m\pi}{a}x\right) e^{j(\omega t - \beta z)} \end{aligned}\right\} \text{TE waves}$$

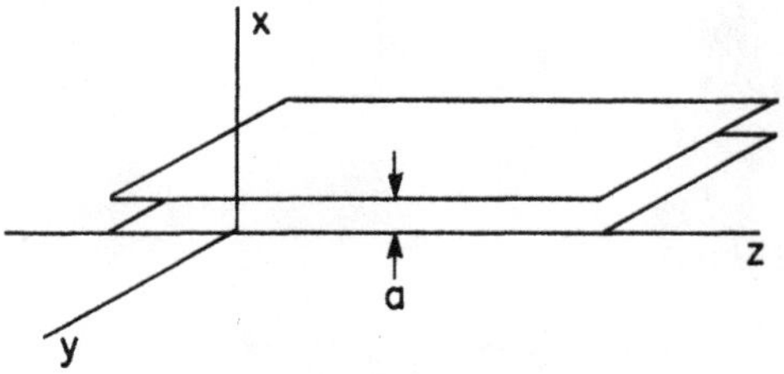

Figure 3.1 Parallel conducting planes.

$$\left.\begin{aligned} E_x &= \frac{\beta}{\omega\epsilon} C_2 \cos\left(\frac{m\pi}{a}x\right) e^{j(\omega t - \beta z)} \\ E_z &= \frac{jm\pi}{\omega\epsilon a} C_2 \sin\left(\frac{m\pi}{a}x\right) e^{j(\omega t - \beta z)} \\ H_y &= C_2 \cos\left(\frac{m\pi}{a}x\right) e^{j(\omega t - \beta z)} \end{aligned}\right\} \text{TM waves}$$

2. Propagation Constant:

$$\gamma = \sqrt{\left(\frac{m\pi}{a}\right)^2 - \omega^2\mu\epsilon}$$

3. Critical Frequency:

$$f_c = \frac{\omega_c}{2\pi} = \frac{m}{2a\sqrt{\mu\epsilon}}$$

4. Cutoff Free-Space Wavelength:

$$\lambda_c = \frac{c}{f_c} = \frac{2a}{m} c\sqrt{\mu\epsilon}$$

5. Phase-Shift Constant for $f > f_c$:

$$\beta = \frac{2\pi}{\lambda_g} = \sqrt{\omega^2\mu\epsilon - \left(\frac{m\pi}{a}\right)^2}$$

6. Attenuation Constant for $f < f_c$:

$$\alpha = \sqrt{\left(\frac{m\pi}{a}\right)^2 - \omega^2\mu\epsilon}$$

7. Guide Wavelength:

$$\lambda_g = \frac{2\pi}{\beta} = \frac{2\pi}{\sqrt{\omega^2\mu\epsilon - \left(\frac{m\pi}{a}\right)^2}}$$

and in vacuum dielectric:

$$\lambda_g = \frac{\lambda_0}{\sqrt{1 - \left(\frac{m\lambda_0}{2a}\right)^2}}$$

8. Wave Velocity:

$$v = \lambda_g f = \frac{\omega}{\beta} = \frac{\omega}{\sqrt{\omega^2\mu\epsilon - \left(\frac{m\pi}{a}\right)^2}}$$

9. Attenuation Caused by Finite Conductivity:

$$\alpha = \frac{\omega\epsilon\mathcal{R}_s}{a\beta} = \frac{\omega\epsilon\sqrt{\frac{\omega\mu_m}{2\sigma_m}}}{a\sqrt{\omega^2\mu\epsilon - \left(\frac{m\pi}{a}\right)^2}} \quad \text{TM waves}$$

$$\alpha = \frac{2m^2\pi^2\mathcal{R}_s}{\omega\epsilon a^3\beta} = \frac{2m^2\pi^2\sqrt{\frac{\omega\mu_m}{2\sigma_m}}}{\omega\mu a^3\sqrt{\omega^2\mu\epsilon - \left(\frac{m\pi}{a}\right)^2}} \quad \text{TE waves}$$

3.1.2 TEM Modes

1. Field Equations:

$$E_x = \frac{\beta}{\omega\epsilon} C_3 \, e^{j(\omega t - \beta z)}$$

$$H_y = C_3 \, e^{j(\omega t - \beta z)}$$

2. Propagation Constant:

$$\gamma = j\omega\sqrt{\mu\epsilon}$$

3. Phase-Shift Constant:

$$\beta = \frac{2\pi}{\lambda_g} = \omega\sqrt{\mu\epsilon}$$

4. Guide Wavelength:

$$\lambda_g = \frac{2\pi}{\beta} = \frac{2\pi}{\omega\sqrt{\mu\epsilon}}$$

and in vacuum dielectric:

$$\lambda_g = \frac{c}{f}$$

5. Wave Velocity:

$$v = \frac{1}{\sqrt{\mu\epsilon}}$$

6. Attenuation Caused by Finite Conductivity:

$$\alpha = \frac{\mathcal{R}_s}{\eta a} = \frac{\sqrt{\dfrac{\omega\mu_m}{2\sigma_m}}}{\sqrt{\dfrac{\mu}{\epsilon}}\, a}$$

7. Characteristic Impedance of TEM Mode:
voltage-current basis

$$Z_0 = \frac{V}{I} = \sqrt{\frac{\mu}{\epsilon}}\frac{a}{y_0} = \eta\frac{a}{y_0}$$

power basis

$$Z_0 = \frac{P}{I_{\text{rms}}^2} = \sqrt{\frac{\mu}{\epsilon}}\frac{a}{y_0} = \eta\frac{a}{y_0}$$

wave impedance

$$Z_0 = \frac{E_x}{H_y} = \sqrt{\frac{\mu}{\epsilon}} = \eta$$

where y_0 is the width of a segment of the parallel plates under consideration.

3.2 WAVES IN RECTANGULAR GUIDES

Consider a perfectly conducting cylinder of rectangular cross section and of infinite extent in the z direction with a nonconducting interior region as illustrated in Figure 3.2. We assume that propagation is in the z direction and that the variation of all field components in this direction may be expressed in the form

$$e^{-\gamma z}$$

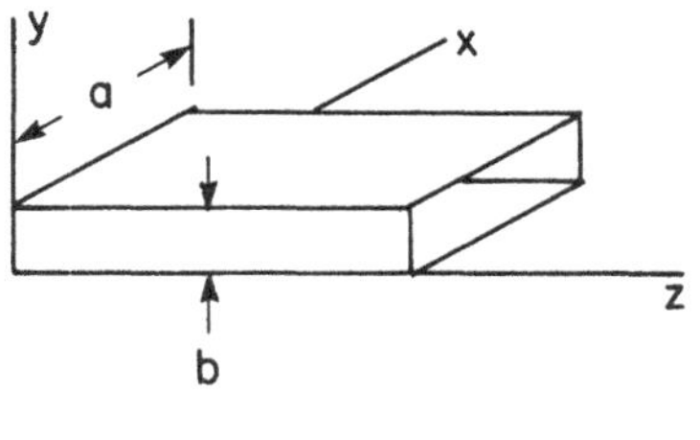

Figure 3.2 Rectangular waveguide.

where, in general,

$$\gamma = \alpha + j\beta$$

is a complex propagation constant.

3.2.1 TE and TM Modes

1. Field Equations:

$$\left.\begin{aligned}
E_x &= \frac{j\omega\mu}{h^2} C_1 B \cos Ax \sin By \\
E_y &= \frac{-j\omega\mu}{h^2} C_1 A \sin Ax \cos By \\
H_x &= \frac{j\beta}{h^2} C_1 A \sin Ax \cos By \\
H_y &= \frac{j\beta}{h^2} C_1 B \cos Ax \sin By \\
H_z &= C_1 \cos Ax \cos By
\end{aligned}\right\} \text{TE}_{mn} \text{ waves}$$

$$\left.\begin{aligned}
E_x &= \frac{-j\beta}{h^2} C_2 A \cos Ax \sin By \\
E_y &= \frac{-j\beta}{h^2} C_2 B \sin Ax \cos By \\
E_z &= C_2 \sin Ax \sin By \\
H_x &= \frac{j\omega\epsilon}{h^2} C_2 B \sin Ax \cos By \\
H_y &= \frac{-j\omega\epsilon}{h^2} C_2 A \cos Ax \sin By
\end{aligned}\right\} \text{TM}_{mn} \text{ waves}$$

where

$$A = \frac{m\pi}{a} \quad B = \frac{n\pi}{b}$$

$$h^2 = \gamma^2 + \omega^2\mu\epsilon$$

and each component is multiplied by the expression $e^{j(\omega t - \beta z)}$.

2. Propagation Constant:

$$\gamma = \sqrt{h^2 - \omega^2\mu\epsilon} = \sqrt{A^2 + B^2 - \omega^2\mu\epsilon}$$
$$= \sqrt{\left(\frac{m\pi}{a}\right)^2 + \left(\frac{n\pi}{b}\right)^2 - \omega^2\mu\epsilon}$$

3. Critical Frequency:

$$f_c = \frac{1}{2\pi\sqrt{\mu\epsilon}}\sqrt{\left(\frac{m\pi}{a}\right)^2 + \left(\frac{n\pi}{b}\right)^2}$$

4. Cutoff Wavelength (measured in the dielectric medium):

$$\lambda_c = \frac{2}{\sqrt{\left(\frac{m}{a}\right)^2 + \left(\frac{n}{b}\right)^2}}$$

5. Phase-Shift Constant for $f > f_c$:

$$\beta = \sqrt{\omega^2\mu\epsilon - \left(\frac{m\pi}{a}\right)^2 - \left(\frac{n\pi}{b}\right)^2}$$

6. Attenuation Constant for $f < f_c$:

$$\alpha = \sqrt{\left(\frac{m\pi}{a}\right)^2 + \left(\frac{n\pi}{b}\right)^2 - \omega^2\mu\epsilon}$$

7. Guide Wavelength:

$$\lambda_g = \frac{2\pi}{\beta} = \frac{2\pi}{\sqrt{\omega^2\mu\epsilon - \left(\frac{m\pi}{a}\right)^2 - \left(\frac{n\pi}{b}\right)^2}}$$

and in vacuum dielectric:

$$\lambda_g = \frac{\lambda_0}{\sqrt{1 - \left(\frac{m\lambda_0}{2a}\right)^2 - \left(\frac{n\lambda_0}{2b}\right)^2}}$$

8. Wave Velocity:

$$v = \lambda_g f = \frac{\omega}{\beta} = \frac{\omega}{\sqrt{\omega^2\mu\epsilon - \left(\frac{m\pi}{a}\right)^2 - \left(\frac{n\pi}{b}\right)^2}}$$

9. Attenuation of TE_{10} Mode in Vacuum-Filled Waveguide Caused by Finite Conductivity of the Guide Walls:

$$\alpha = \frac{\mathcal{R}_s}{\eta_0 b} \frac{1 + \frac{2b}{a}\left(\frac{\lambda_0}{2a}\right)^2}{\sqrt{1 - \left(\frac{\lambda_0}{2a}\right)^2}}$$

where

$$\mathcal{R}_s = \sqrt{\frac{\omega\mu_m}{2\sigma_m}} \qquad \eta_0 = \sqrt{\frac{\mu_0}{\epsilon_0}}$$

10. Characteristic Impedance of the TE_{10} Mode in Vacuum Waveguide:
voltage-current basis

$$Z_0 = \frac{\pi\eta_0}{2} \frac{b}{a\sqrt{1 - \left(\frac{\lambda_0}{2a}\right)^2}}$$

power-current basis

$$Z_0 = \frac{\pi^2\eta_0}{8} \frac{b}{a\sqrt{1 - \left(\frac{\lambda_0}{2a}\right)^2}}$$

power-voltage basis

$$Z_0 = 2\eta_0 \frac{b}{a\sqrt{1 - \left(\frac{\lambda_0}{2a}\right)^2}}$$

wave impedance

$$Z_0 = \frac{-E_y}{H_x} = \frac{\eta_0}{\sqrt{1 - \left(\frac{\lambda_0}{2a}\right)^2}}$$

where

$$\lambda_0 = \frac{1}{f\sqrt{\mu_0 \epsilon_0}}$$

3.3 SURFACE CHARACTERISTICS

1. Surface Impedance (of good conductors if the conductor is very much thicker than the current skin depth):

$$Z_s = \sqrt{\frac{j\omega\mu_m}{\sigma_m}}$$

2. Surface Resistance:

$$\mathcal{R}_s = \sqrt{\frac{\omega\mu_m}{2\sigma_m}}$$

3. Surface Reactance:

$$X_s = \sqrt{\frac{\omega\mu_m}{2\sigma_m}}$$

4. Skin Depth (of currents):

$$\delta = \sqrt{\frac{2}{\omega\mu_m\sigma_m}}$$

3.4 TWO-WIRE TRANSMISSION LINES (see Figure 3.3)

1. Characteristic Impedance:

$$Z_0 = \frac{120}{\sqrt{\epsilon_1}} \cosh^{-1} \frac{D}{d} \qquad (\Omega)$$

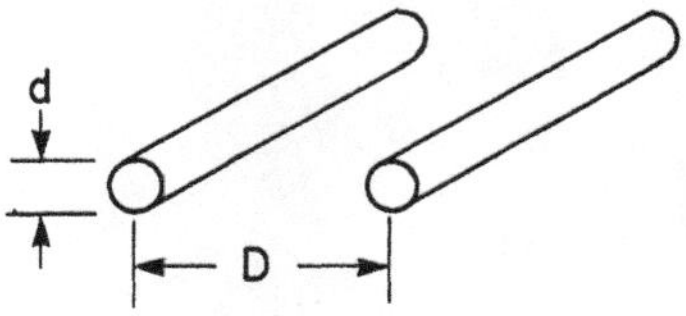

Figure 3.3 Two-wire transmission line.

3.5 COAXIAL TRANSMISSION LINES (see Figure 3.4)

1. Characteristic Impedance:

$$Z_0 = \frac{138}{\sqrt{\epsilon_1}} \log \frac{b}{a} \qquad (\Omega)$$

2. Cutoff Wavelength (of first higher mode, TE_{11}):

$$\lambda_c = \pi \sqrt{\epsilon_1}\,(a + b) \qquad \text{(within 8\% accuracy [1])}$$

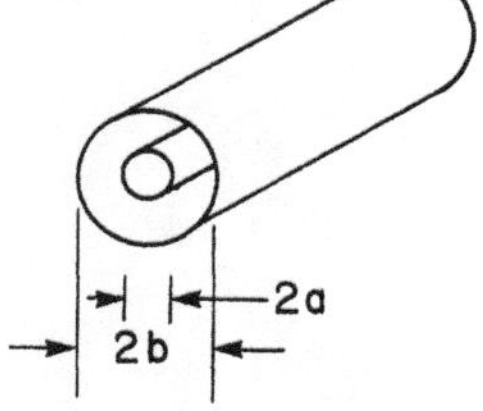

Figure 3.4 Coaxial transmission line.

3.6 CIRCULAR WAVEGUIDES (see Figure 3.5)

1. Cutoff Wavelength:

$$\lambda_c = \sqrt{\mu_1 \epsilon_1}\, D_{mn} a$$

where D_{mn} is given in Table 3.1.

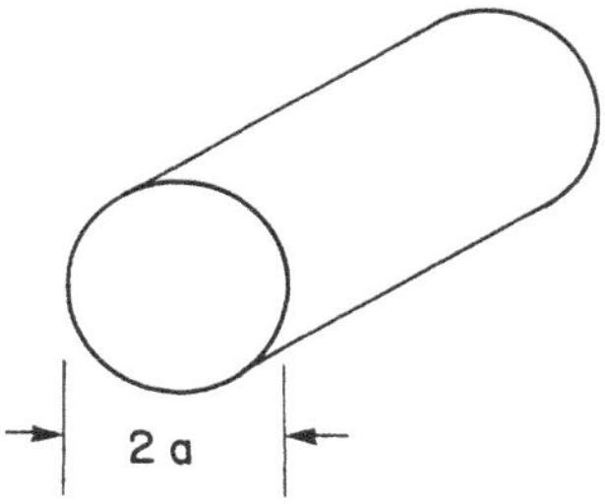

Figure 3.5 Circular waveguide.

Table 3.1
Cutoff Constants for Circular Waveguide*

D_{mn} for TE_{mn} waves					D_{mn} for TM_{mn} waves				
n \ m	0	1	2	3	n \ m	0	1	2	3
1	1.640	3.412	2.057	1.496	1	2.613	1.640	1.224	0.966
2	0.896	1.178	0.937	0.764	2	1.139	0.896	0.747	0.644
3	0.618	0.736	0.631	0.554	3	0.726	0.618	0.541	0.482
4	0.475	0.54	0.48	0.44	4	0.534	0.475	0.425	0.388

*After Loman [3].

3.7 PROPAGATION MODES

Illustrations of several propagating modes in rectangular guides, circular guides, and coaxial lines are shown in Figures 3.6 through 3.11. (In Figures 3.10 and 3.11 the definitions of a and b are reversed from those indicated in Figure 3.4.)

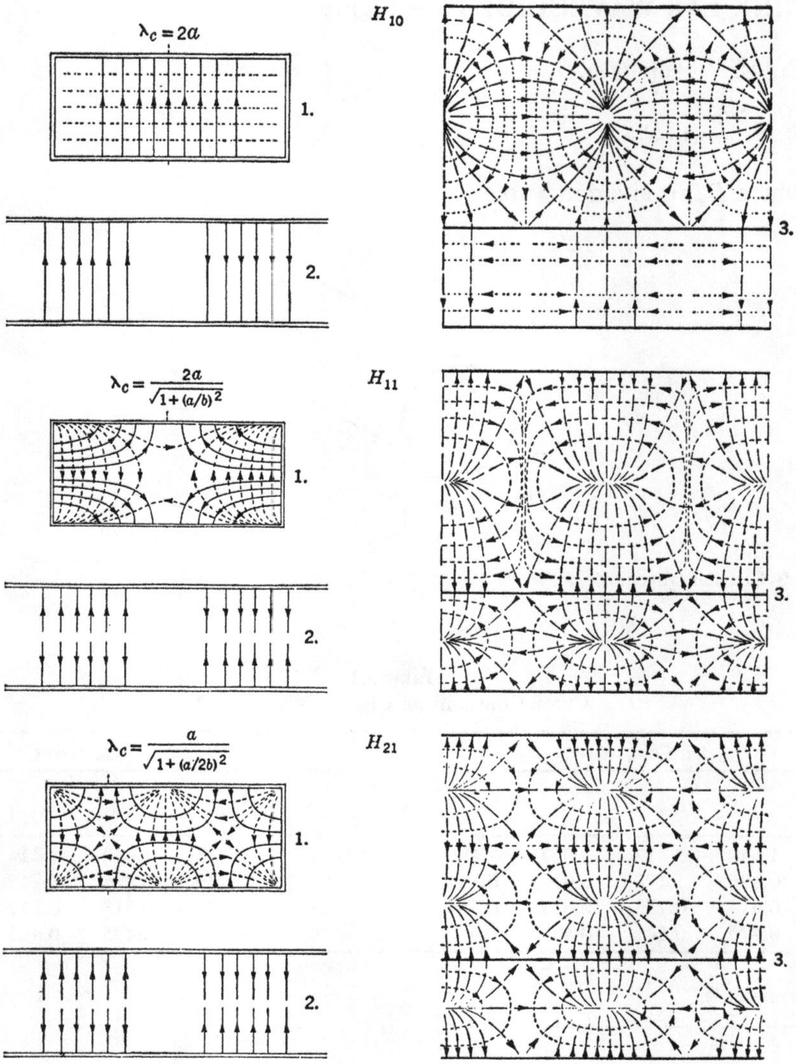

Figure 3.6 Field distribution for TE modes (or H modes) in rectangular waveguide. (From Marcuvitz [2].)

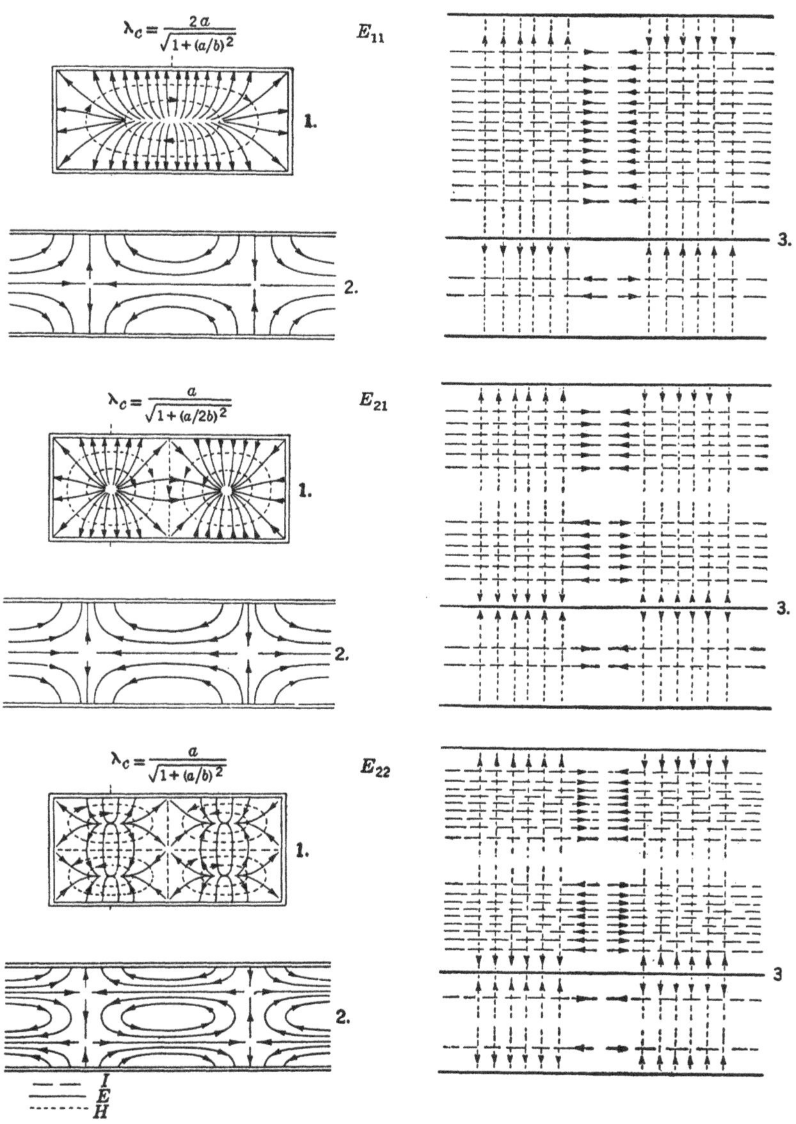

Figure 3.7 Field distribution for TM modes (or E modes) in rectangular waveguide. (From Marcuvitz [2].)

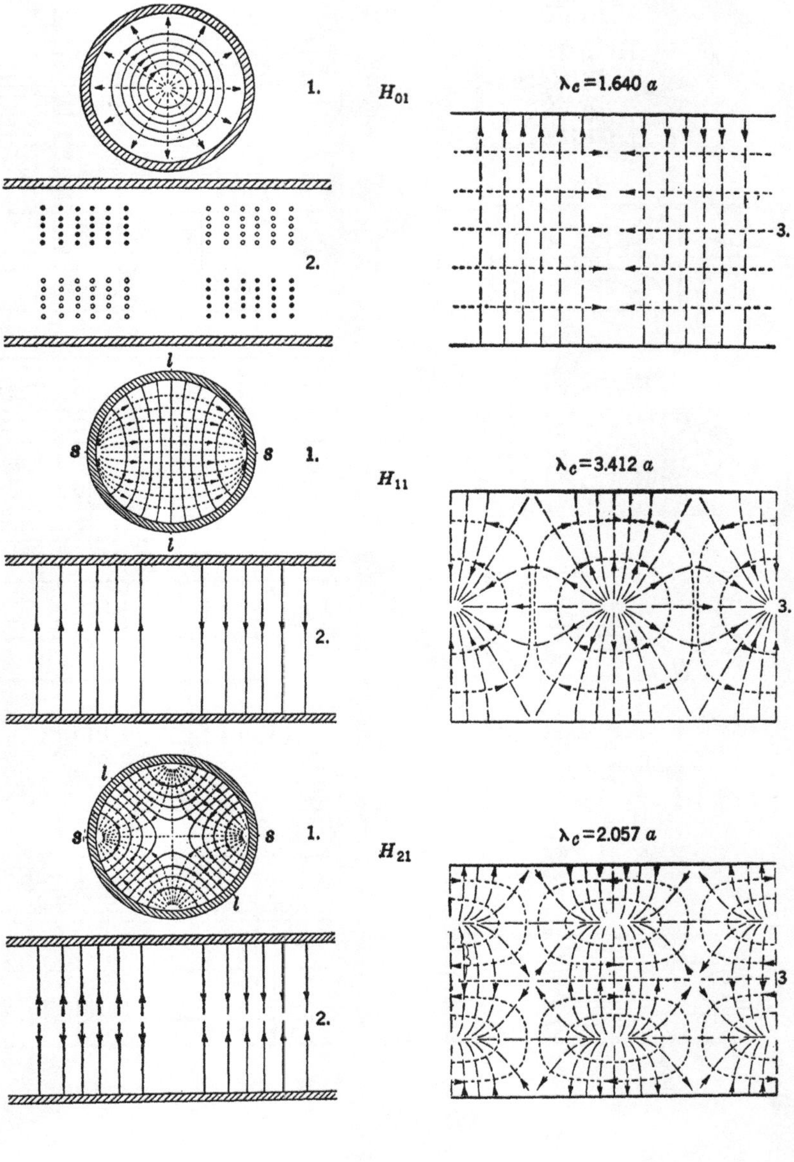

Figure 3.8 Field distribution for TE modes (or H modes) in circular waveguide. (From Marcuvitz [2].)

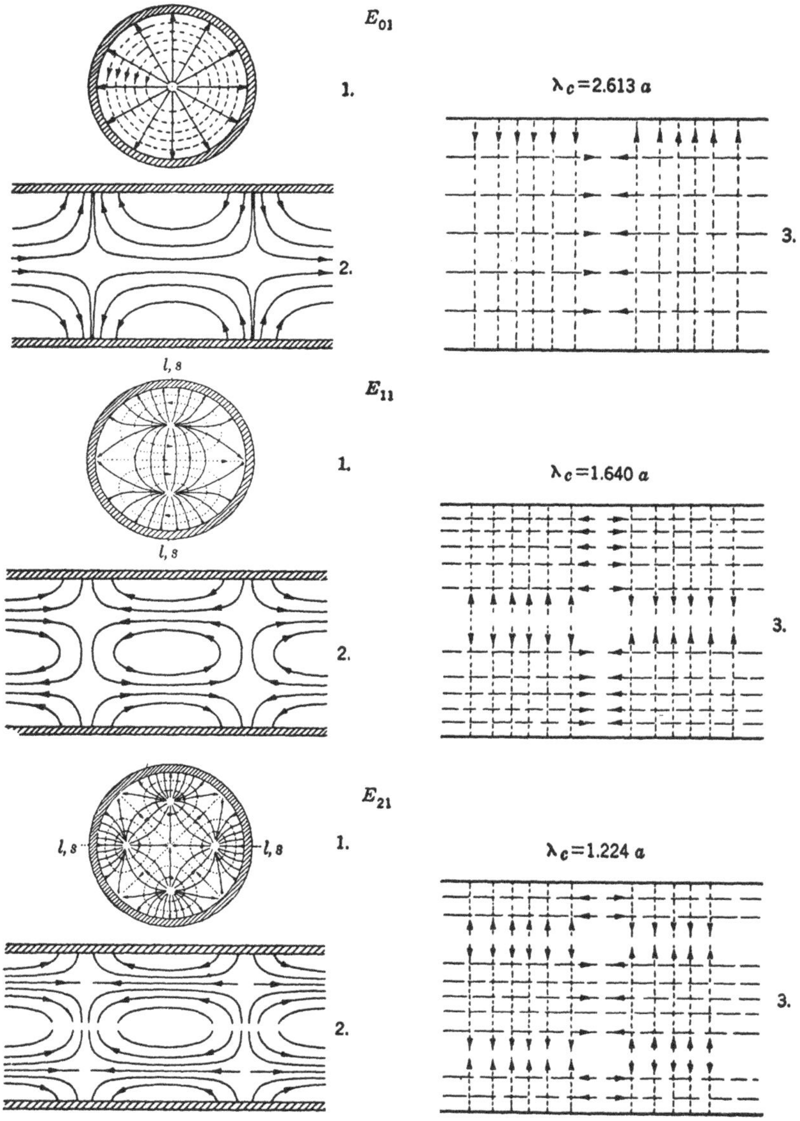

Figure 3.9 Field distribution for TM modes (or E modes) in circular waveguide. (From Marcuvitz [2].)

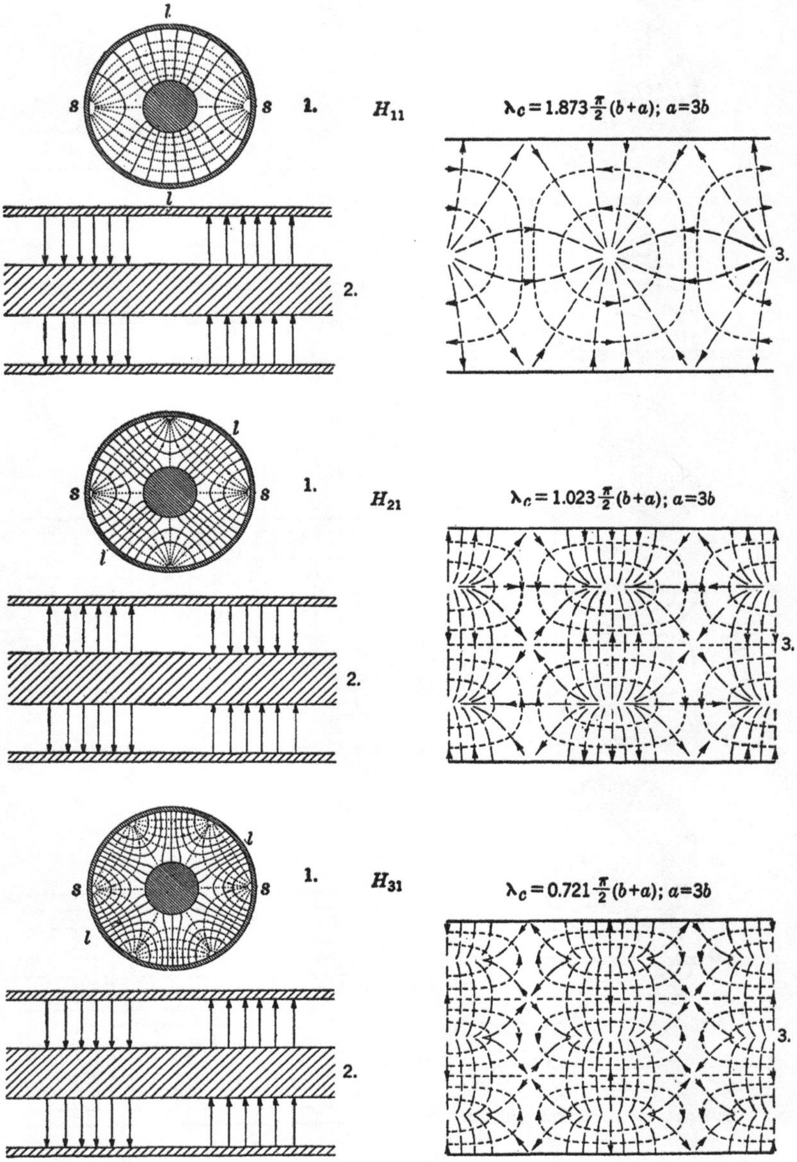

Figure 3.10 Field distribution for TE modes (or H modes) in coaxial waveguide. (From Marcuvitz [2].)

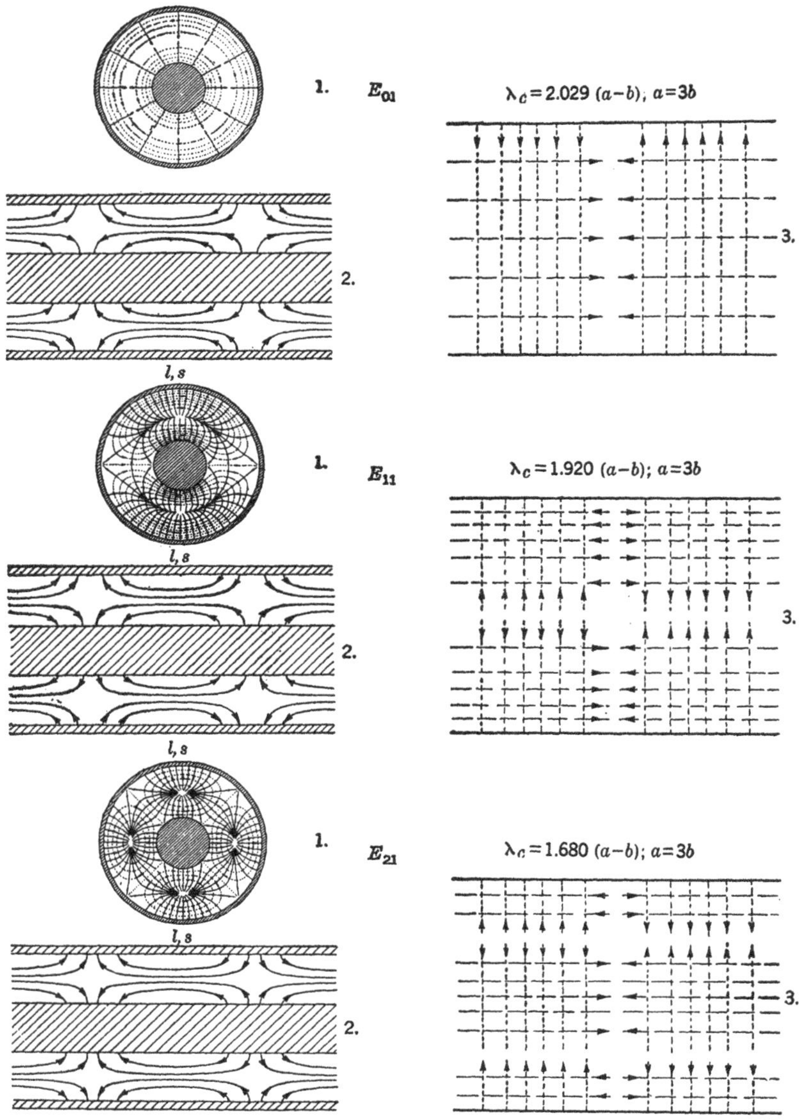

Figure 3.11 Field distribution for TM modes (or E modes) in coaxial waveguide. (From Marcuvitz [2].)

REFERENCES

1. T. Moreno, *Microwave Transmission Design Data* (originally published by Sperry Gyroscope Company, 1948; reprinted by New York: Dover Publications, 1958; republished by Artech House, 1989).
2. N. Marcuvitz, *Waveguide Handbook,* MIT Radiation Laboratory Series, Vol. 10, New York: McGraw-Hill, 1951.
3. R.V. Lowman, "Transmission Lines and Waveguides," Chapter 42 in *Antenna Engineering Handbook,* 2nd Ed. (R.C. Johnson and H. Jasik, eds.), New York: McGraw-Hill, 1984.
4. R.I. Sarbacher and W.A. Edson, *Hyper and Ultrahigh Frequency Engineering,* New York: John Wiley and Sons, 1943.
5. E.C. Jordan, *Electromagnetic Waves and Radiating Systems,* New York: Prentice-Hall, 1950.
6. S. Kuhn, "Calculation of Attenuation in Waveguides," *J. Inst. Elec. Eng.*, Vol. 93, Part IIIA, No. 4, 1946, p. 663.

LIST OF SYMBOLS

A	$m\pi/a$
a	separation of parallel-plate waveguide, wide dimension of rectangular waveguide, radius of inner conductor of coaxial line, or radius of circular waveguide
B	$\mathrm{n}\pi/b$
b	narrow dimension of rectangular waveguide or radius of outer conductor of coaxial line
C_1, C_2, C_3	arbitrary constants
c	velocity of light in free space
D	separation of wires in two-wire transmission line
d	diameter of wire in two-wire transmission line
E_x, E_y, E_z	components of the electric field in Cartesian coordinates
f	wave frequency
f_c	critical (cutoff) frequency
H_x, H_y, H_z	components of the magnetic field in Cartesian coordinates
h^2	$\gamma^2 + \omega^2 \mu\epsilon$
j	imaginary symbol, $j = \sqrt{-1}$
m, n	integers
$\mathcal{R}_s$	surface resistance
t	time variable
v	wave velocity
x, y, z	Cartesian coordinate variables
y_0	width of a segment of parallel plates under consideration
Z_0	characteristic impedance
Z_s	surface impedance

α	value of the real part of the complex propagation constant or attenuation constant caused by the finite conductivity of the waveguide walls
β	value of the imaginary part of the complex propagation constant
γ	complex propagation constant
δ	skin depth of currents
ϵ	permittivity
ϵ_1	relative permittivity
η	intrinsic impedance
η_0	intrinsic impedance of free space
λ	wavelength in medium
λ_c	critical (cutoff) wavelength
λ_g	guide wavelength
λ_0	wavelength in free space
μ	permeability
μ_1	relative permeability
μ_m	permeability of the conducting walls
σ_m	conductivity of the conducting walls
ω	angular frequency, rad/s
ω_c	critical (cutoff) angular frequency

Chapter 4
HORNS

Horns [1–4] are probably the most common type of microwave antenna. They have been developed in a wide variety of configurations; the most popular types will be presented here.

4.1 RECTANGULAR HORNS

4.1.1 Patterns

The general configuration of rectangular horns is illustrated in Figure 4.1. We assume that the horn is excited at the throat by the dominant TE_{10} mode in rectangular waveguide. In the aperture, the field amplitude is assumed to have the same distribution as that of the exciting mode (uniform in the E-plane and cosine in the H-plane), and the wavefront has radii of curvatures equal to R_e and R_h.

The normalized path errors are (see Section A.16 in the Appendix):

$$s_e = \frac{\Delta_e}{\lambda} \approx \frac{B^2}{8\lambda R_e} \qquad s_h = \frac{\Delta_h}{\lambda} \approx \frac{A^2}{8\lambda R_h} \tag{4.1}$$

In the aperture, the electric field can be approximated by

$$E_y = E_0 \cos\left(\frac{\pi x}{A}\right) \exp\left\{-j2\pi\left[s_e\left(\frac{2y}{B}\right)^2 + s_h\left(\frac{2x}{A}\right)^2\right]\right\} \tag{4.2}$$

where the rectangular xy coordinate system is centered in the aperture plane (p. 181 of [3]).

Calculated E- and H-plane patterns for various normalized path errors are presented in Figures 4.2 and 4.3. These patterns are reasonably accurate when the aperture dimensions are several wavelengths. For smaller horns, the pattern ampli-

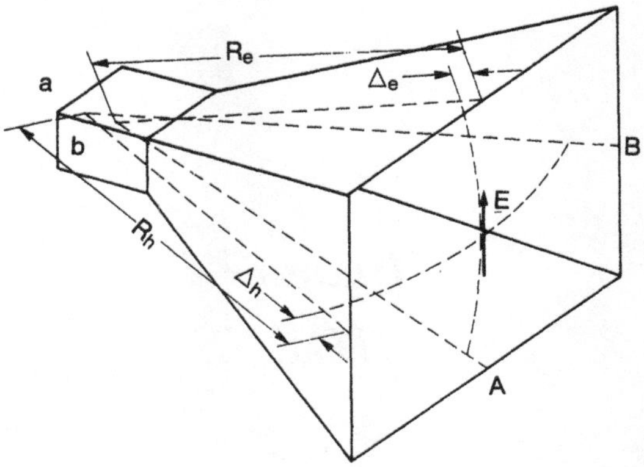

Figure 4.1 General configuration of rectangular horns. (After Milligan [3], with permission of McGraw-Hill, Inc.)

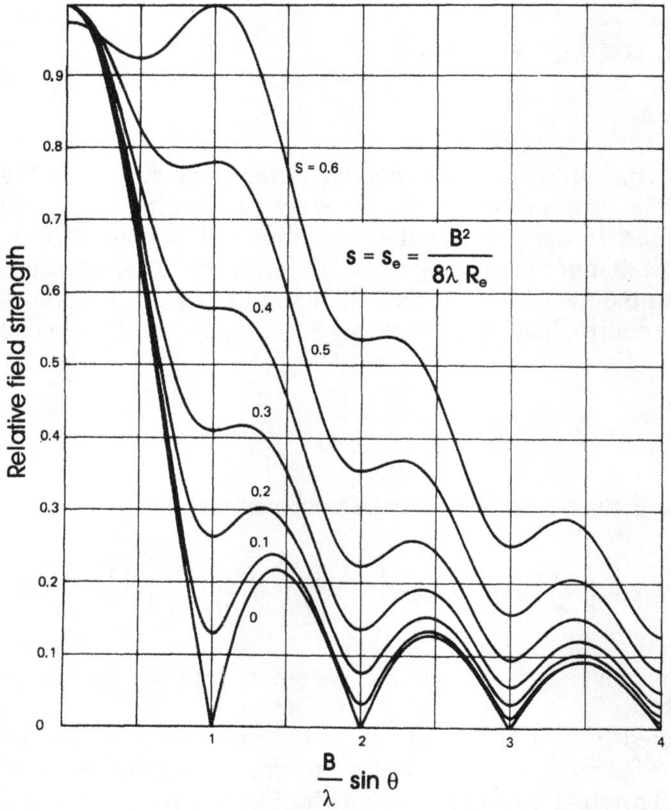

Figure 4.2 Universal E-plane patterns of rectangular horns for various normalized path errors and TE_{10} mode excitation. (After Milligan [3], with permission of McGraw-Hill, Inc.)

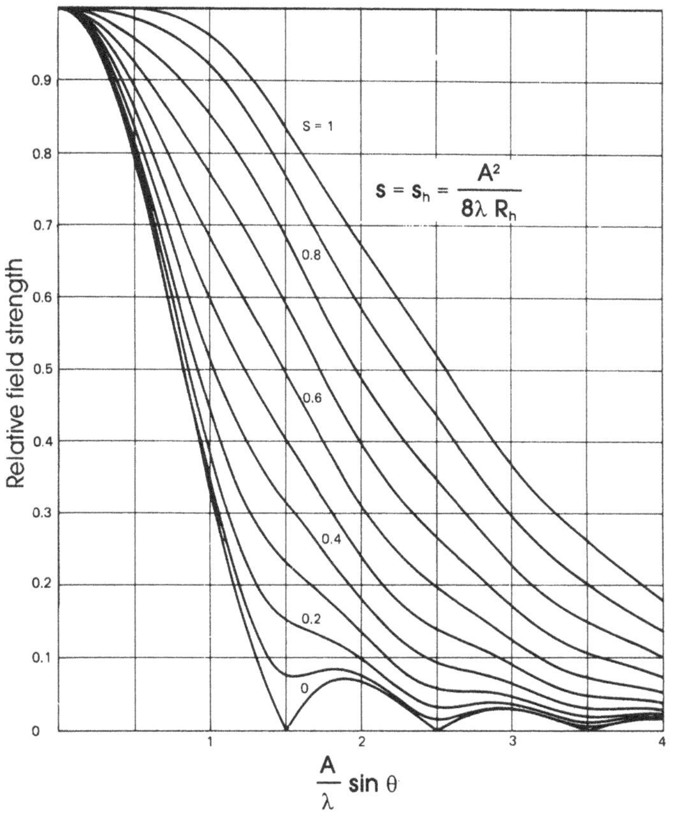

Figure 4.3 Universal H-plane patterns of rectangular horns for various normalized path errors and TE_{10} mode excitation. (After Milligan [3], with permission of McGraw-Hill, Inc.)

tude must be multiplied by the appropriate obliquity factor. The approximate obliquity factors,

$$F_e \approx F_h \approx \frac{1 + \cos\theta}{2} \tag{4.3}$$

where θ is the pattern angle, are reasonably accurate for aperture sizes greater than 1.5 wavelengths.

The calculation of accurate patterns from small horns is very difficult because currents on the external surfaces of the horn and reflections at the aperture influence the radiation pattern. Fortunately, we can refer to experimental data presented by Risser [2]; the experimental 10-dB widths of small horns are presented in Figure 4.4. All of the horns had small phase variations over their apertures.

Refer to Figure 4.5 for an illustration of phase errors at the aperture of a flared horn. It can be shown (see Section A.16) that

$$\frac{\Delta}{\lambda_g} \approx \frac{D^2}{8R\lambda_g} \quad \text{or} \quad \frac{\Delta}{\lambda_g} \approx \frac{D \sin\theta_f}{4\lambda_g} \tag{4.4}$$

Using 1/8 as the allowable upper limit for Δ/λ_g,

$$R > \frac{D^2}{\lambda_g} \quad \text{or} \quad \theta_f < \sin^{-1}\left(\frac{\lambda_g}{2D}\right) \tag{4.5}$$

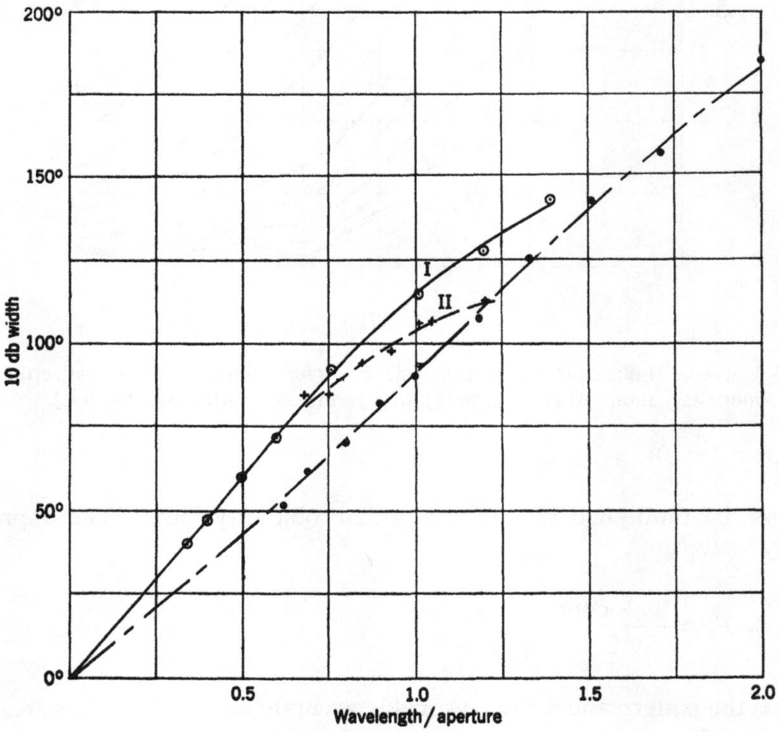

Figure 4.4 Experimental 10-dB widths of horns having small phase variations over the aperture. — - — E-plane; — — — H-plane sectoral horns; ———— H-plane compound horns with E-plane aperture equal to or greater than a wavelength. Compound horns have flares in both planes. (From Risser [2].)

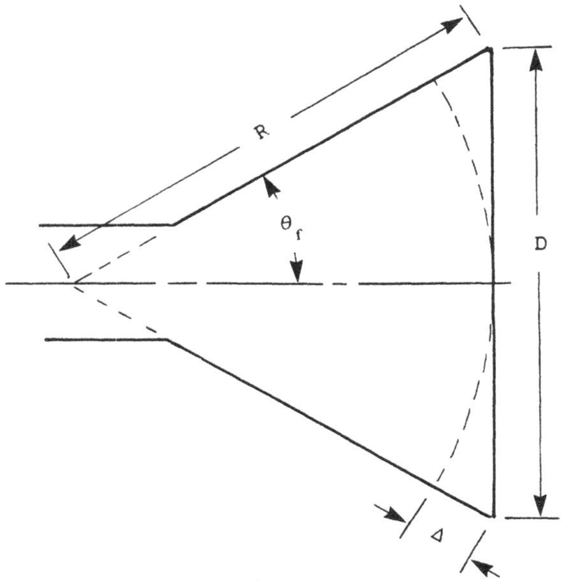

Figure 4.5 Illustration of phase error in the aperture of a flared horn.

Empirical formulas were presented by Risser for the 10-dB beamwidth as a function of aperture size for the average small horn.

1. For the electric plane,

$$\theta_{E10} \approx \frac{88\lambda}{B} \quad \text{(degrees)} \qquad \text{for } \frac{B}{\lambda} < 2.5 \tag{4.6}$$

2. For the magnetic plane

$$\theta_{H10} \approx 31 + \frac{79\lambda}{A} \quad \text{(degrees)} \qquad \text{for } \frac{A}{\lambda} < 3 \tag{4.7}$$

These formulas do not predict the 10-dB beamwidths with great accuracy, but they are useful as a first approximation when designing small horns.

For the convenience of horn designers, the data of Figure 4.4 have been plotted in larger graphs with more grid lines in Figures 4.6 and 4.7.

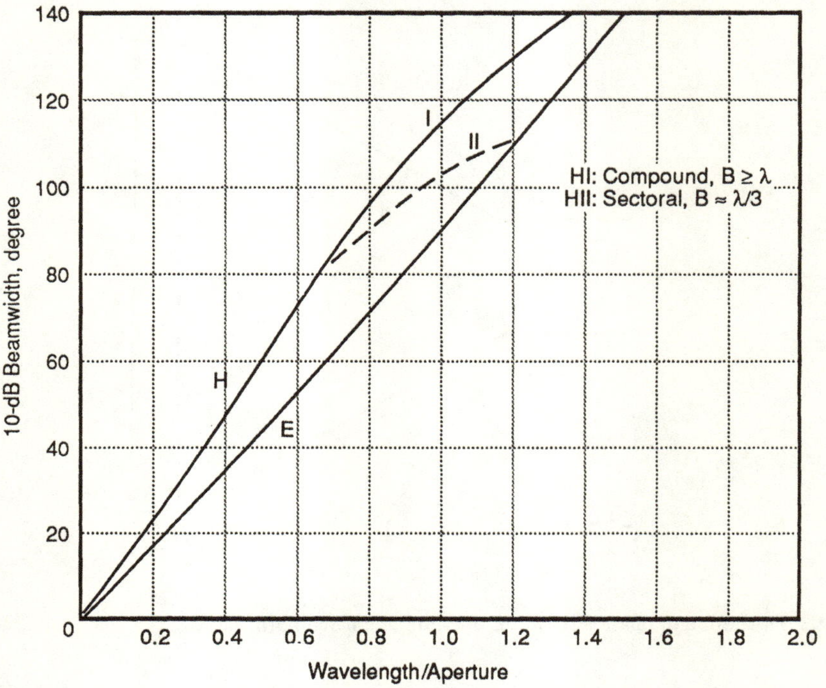

Figure 4.6 Experimental data of Figure 4.4 expanded with more grid lines.

4.1.2 Directivity and Gain

For most microwave horns, the losses are small, so the gain and directivity are essentially identical. At the higher frequencies, however, we may need to account for dissipative losses in the horn. In this section, we will assume that gain and directivity are equal.

The gain of both sectoral and rectangular horns can be expressed as

$$G = G_0 K_e K_h \tag{4.8}$$

where the area gain (for uniform E-plane and cosine H-plane field distributions) is

$$G_0 = \frac{32AB}{\pi\lambda^2} \tag{4.9}$$

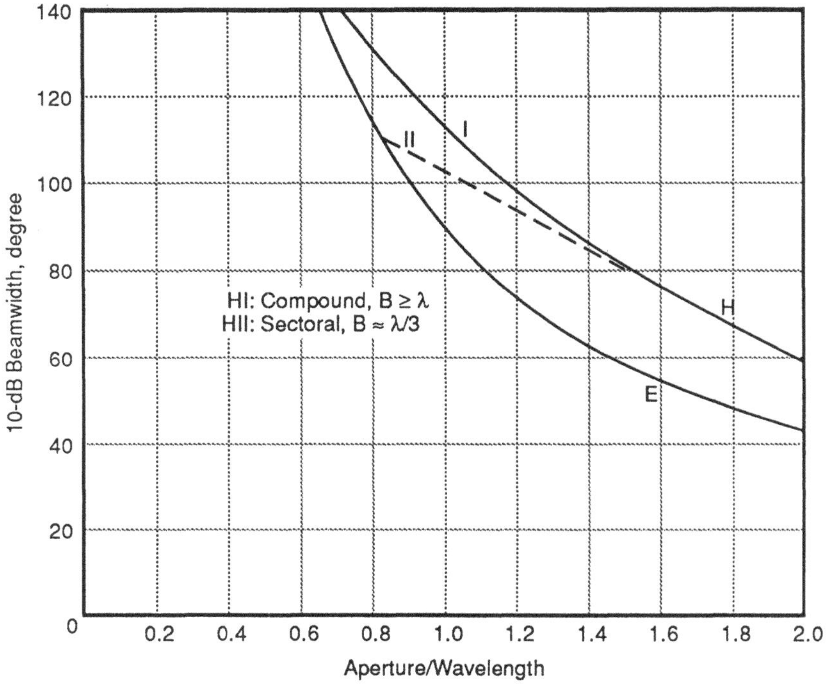

Figure 4.7 Experimental data of Figure 4.4 expanded with more grid lines and beamwidths *versus* aperture/wavelength.

relative to an isotropic radiator. The factors K_e and K_h, both equal to or less than unity, account for the reduction in gain due to phase errors caused by the horn flare. The factor K_e is a function of s_e, and K_h is a function of s_h, where the values of s_e and s_h are given in eq. (4.1).

Converting eq. (4.8) to a decibel form, we have

$$\text{gain} \quad (\text{dB}) = 10.08 + 10 \log \frac{AB}{\lambda^2} - L_e - L_h \tag{4.10}$$

where

$$L_e = -10 \log K_e \qquad L_h = -10 \log K_h \tag{4.11}$$

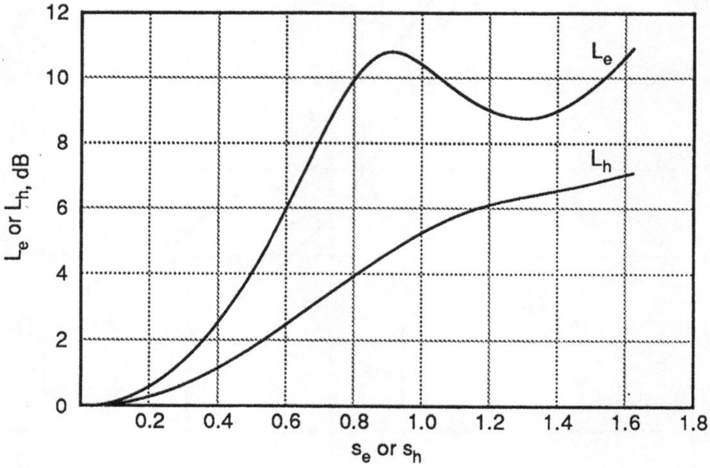

Figure 4.8 Gain-reduction factors for E- and H-plane flares. (Data from Table 4.1.)

Table 4.1
Gain-Reduction Factors, L_e and L_h, for Sectoral and Rectangular Horns*

16s_e or 16s_h	*L_e (dB)*	*L_h (dB)*	*16s_e or 16s_h*	*L_e (dB)*	*L_h (dB)*
1.0	0.060	0.029	14.0	10.783	4.486
1.5	0.134	0.064	15.0	10.849	4.892
2.0	0.239	0.114	16.0	10.502	5.252
2.5	0.374	0.179	18.0	9.474	5.819
3.0	0.541	0.257	20.0	8.847	6.210
3.5	0.738	0.349	22.0	8.901	6.504
4.0	0.967	0.454	24.0	9.637	6.785
4.5	1.229	0.573	26.0	10.938	7.102
5.0	1.525	0.705	28.0	12.430	7.460
5.5	1.854	0.850	30.0	13.312	7.831
6.0	2.218	1.007	32.0	13.052	8.175
6.5	2.618	1.176	34.0	12.251	8.460
7.0	3.054	1.357	36.0	11.666	8.684
7.5	3.527	1.547	38.0	11.607	8.869
8.0	4.037	1.748	40.0	12.121	9.047
9.0	5.166	2.175	42.0	13.104	9.243
10.0	6.427	2.630	44.0	14.221	9.462
11.0	7.769	3.101	46.0	14.851	9.692
12.0	9.081	3.577	48.0	14.619	9.911
13.0	10.163	4.043	50.0	13.937	10.101

*After Love [1] and Jull [5].

The values for L_e and L_h are shown graphically in Figure 4.8 and are tabulated in Table 4.1 as functions of s_e and s_h.

Equation (4.8) includes the geometrical-optics field in the horn aperture and the effects of singly diffracted fields at the aperture edges. It does not include the secondary effects of multiple diffractions at the edges or of reflections within the horn. These secondary effects can cause variations (as a function of frequency) about the gains calculated with eq. (4.8). The variations are on the order of ± 0.5 dB for horns having gains of about 12 dB, ± 0.2 dB for horns having gains of about 18 dB, and less than ± 0.1 dB for horns having gains exceeding 23 dB.

4.1.3 Phase Centers

The phase center is the point from which an antenna appears to radiate a spherical wavefront. Near the axis of the main lobe, the wavefront observed in a particular plane tends to be spherical; however, the wavefront deviates from spherical as the observation angle moves away from the main-lobe axis. In general, the apparent phase center in the E-plane does not coincide with the apparent phase center in the H-plane.

Although most rectangular horns do not have a well-defined phase center, such horns can be designed to radiate a wavefront that is essentially spherical throughout most of the main lobe. Thus, rectangular horns are popular as primary feeds for reflectors and lenses.

A rectangular horn with a quadratic phase error across its aperture will have a phase center behind the aperture (within the horn). The phase-center locations of rectangular horns have been calculated by Muehldorf [6], and his results have been tabulated conveniently by Milligan [3] as shown in Table 4.2. The distance from

Table 4.2
Phase-Center Location of a Rectangular Horn (TE_{10} Mode) behind the Aperture as a Ratio of the Wavefront Radius of Curvature*

s_e or s_h	H-plane L_{ph}/R_h	E-plane L_{pe}/R_e	s_e or s_h	H-plane L_{ph}/R_h	E-plane L_{pe}/R_e
0.00	0.000	0.000	0.28	0.258	0.572
0.04	0.0054	0.011	0.32	0.334	0.755
0.08	0.022	0.045	0.36	0.418	
0.12	0.048	0.102	0.40	0.508	
0.16	0.086	0.182	0.44	0.605	
0.20	0.134	0.286	0.48	0.705	
0.24	0.191	0.416	0.52	0.808	

*After Milligan [3].

the aperture plane to the phase center (see Figure 4.9) divided by the wavefront radius of curvature is presented as a function of the normalized ray-path errors [defined in eq. (4.1)]. The data from Table 4.2 are presented graphically in Figure 4.10.

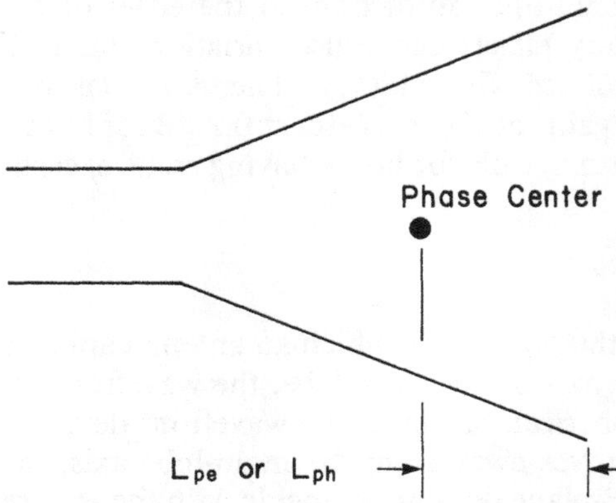

Figure 4.9 Location of phase center in flared horn.

The calculations performed by Muehldorf are valid in the region where the pattern angles are small; thus, the calculated phase-center location should be treated as an approximation (but a good one) for broad-beam horns.

Also, with small horns (compared to a wavelength), currents on the outside surfaces of the horn can influence the location of the phase center. In most cases, the phase center of a small horn is located near and slightly behind the aperture plane; in some cases, however, the phase center may be slightly forward of the aperture plane. A good approximation for small horns is to assume that the phase center is at the aperture plane.

4.2 CIRCULAR-CONICAL HORNS

4.2.1 Patterns

This section will discuss dominant-mode (TE_{11}) circular-conical horns. They are analogous to rectangular horns, which are illustrated in Figure 4.1; however, the feed waveguide is circular, the flared section is a circular cone, and the aperture is circular.

Milligan [3] calculated E- and H-plane patterns for various normalized path errors, and his results are presented in Figures 4.11 and 4.12. He assumed the aper-

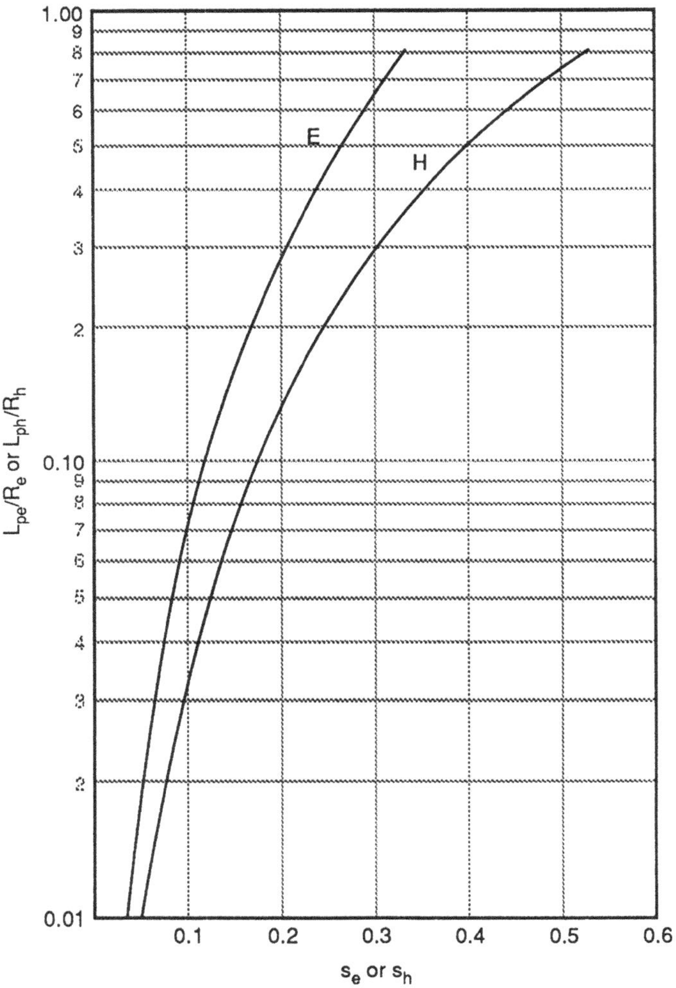

Figure 4.10 Phase-center location behind the aperture of a rectangular horn. (Data from Table 4.2.)

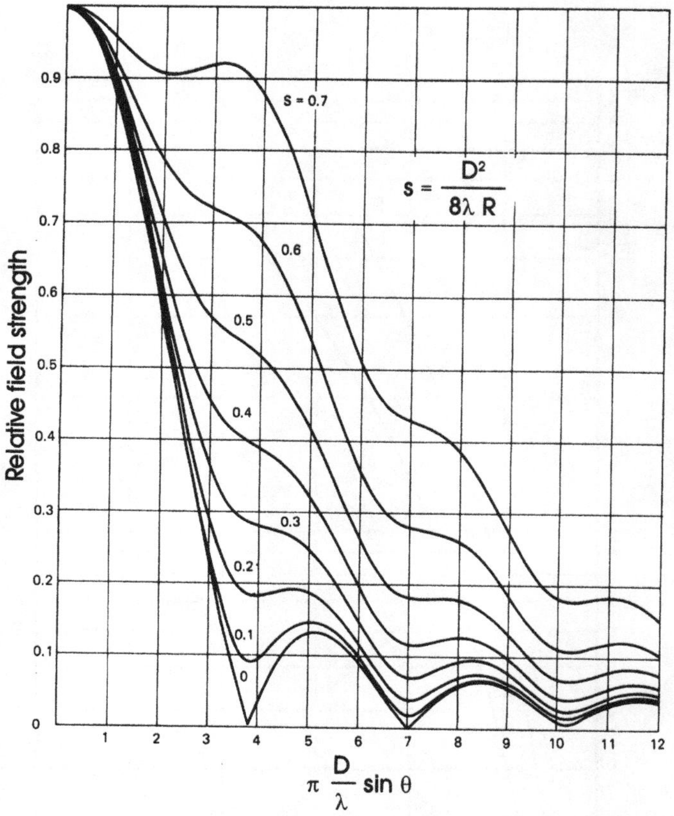

Figure 4.11 Universal E-plane patterns of conical horns for various normalized path errors and TE_{11} mode excitation. (After Milligan [3], with permission of McGraw-Hill, Inc.)

ture amplitude to be that of the TE_{11} mode in circular waveguide and the wavefront radius of curvature to be R, as shown in Figure 4.5.

These theoretical patterns are most accurate when the aperture diameter is several wavelengths. For small horns, the pattern amplitude should be multiplied by the appropriate obliquity factor [see eq. (4.3)].

Experimental patterns of some conical horns were presented by King [1, 7] many years ago. They are reproduced in Figure 4.13 for reference by the reader.

As was the case with rectangular horns, the calculation of accurate patterns for small horns is very difficult because reflections and currents on the external surfaces of the horns affect the radiation patterns. Approximate methods have been applied with some success to open-ended circular waveguide by Chu [2, 8]. Chu's

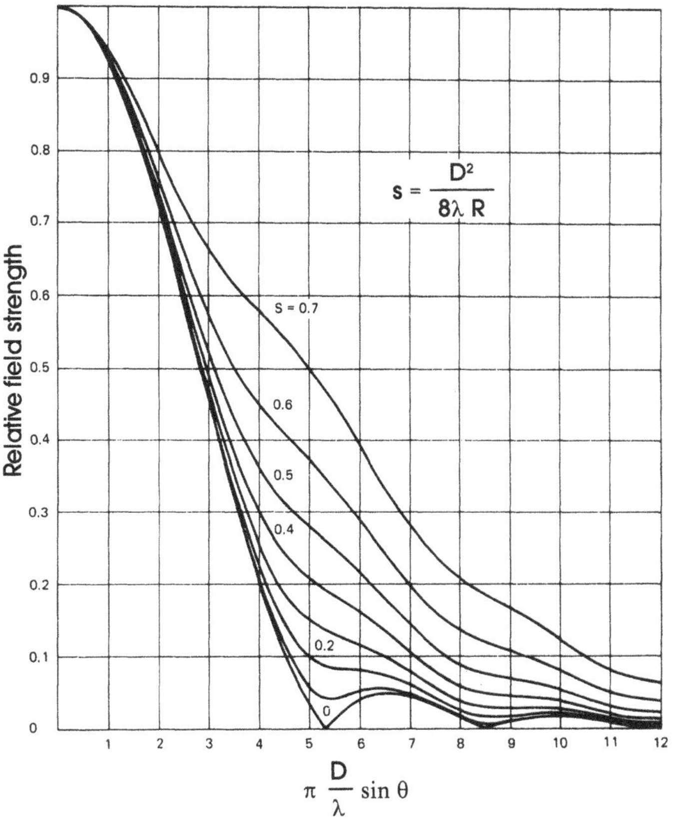

Figure 4.12 Universal H-plane patterns of conical horns for various normalized path errors and TE_{11} mode excitation. (After Milligan [3], with permission of McGraw-Hill, Inc.)

data indicated 3-dB beamwidths *versus* waveguide diameters as indicated in Figure 4.14. For additional information on this subject, see Compton and Collin [9].

4.2.2 Directivity and Gain

Gain estimates for conical horns can be handled in a manner similar to those above for rectangular horns. From Figure 4.5, we have the normalized path error (see Section A.16):

$$s = \frac{\Delta}{\lambda} \approx \frac{D^2}{8\lambda R} \tag{4.12}$$

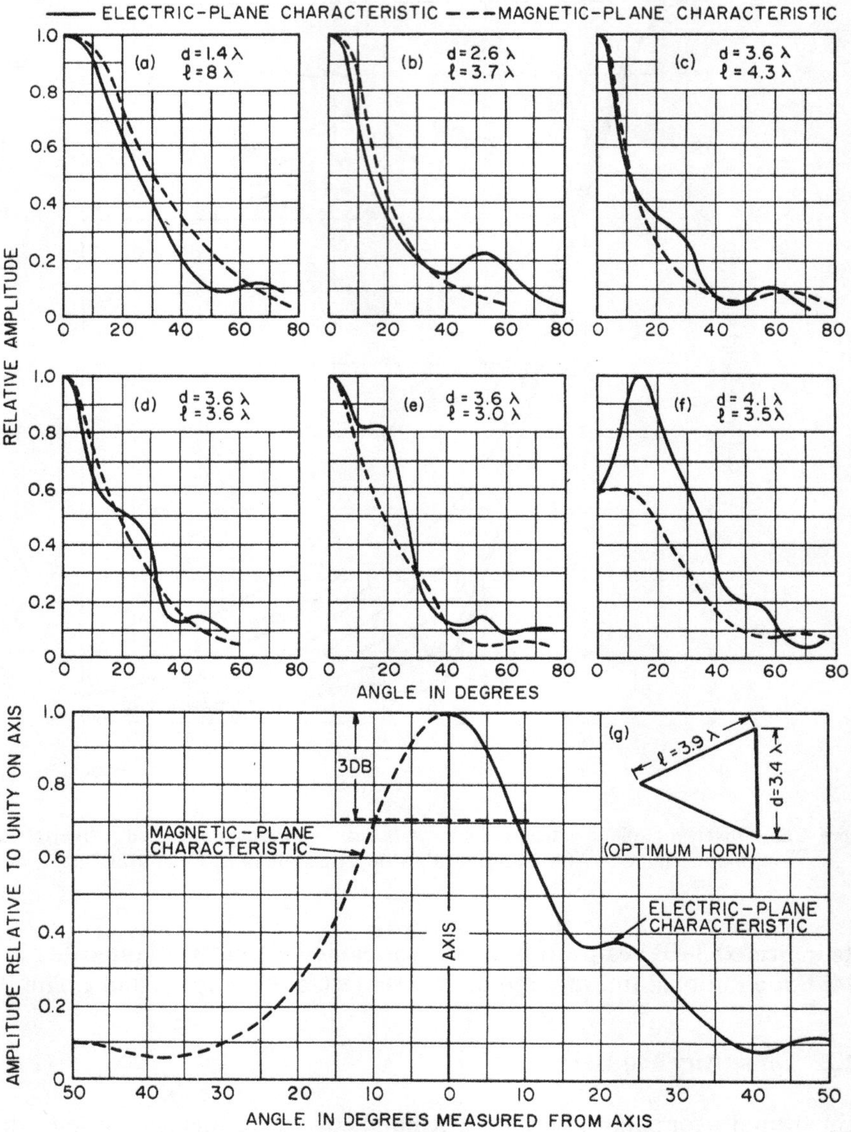

Figure 4.13 Experimentally observed patterns of conical horns of various dimensions. (©1950 IEEE, from King [7].)

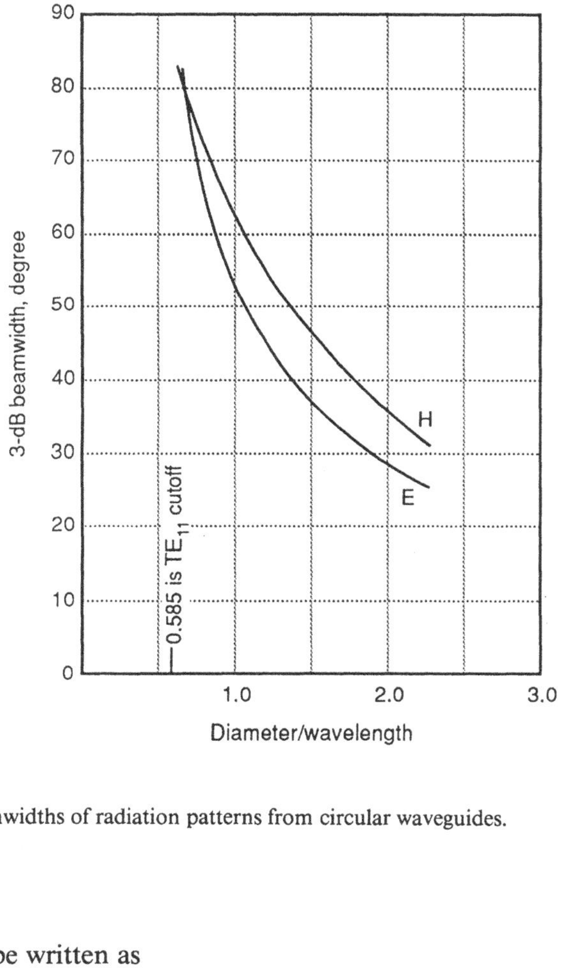

Figure 4.14 Beamwidths of radiation patterns from circular waveguides.

The gain can be written as

$$\text{gain} \quad (\text{dB}) = 20 \log \frac{\pi D}{\lambda} - L_c \tag{4.13}$$

where the first term is the aperture area gain for uniform amplitude and phase, and the second term is the gain loss for both amplitude taper and phase error. Values of the gain-loss term are given in Table 4.3 and in Figure 4.15.

Table 4.3
Gain-Reduction Factor L_c for Circular-Conical Horns*

s	L_c (dB)	s	L_c (dB)	s	L_c (dB)
0.00	0.77	0.28	1.82	0.56	5.28
0.04	0.80	0.32	2.15	0.60	5.98
0.08	0.86	0.36	2.53	0.64	6.79
0.12	0.96	0.40	2.96	0.68	7.66
0.16	1.11	0.44	3.45	0.72	8.62
0.20	1.30	0.48	3.99		
0.24	1.54	0.52	4.59		

*After Milligan [3].

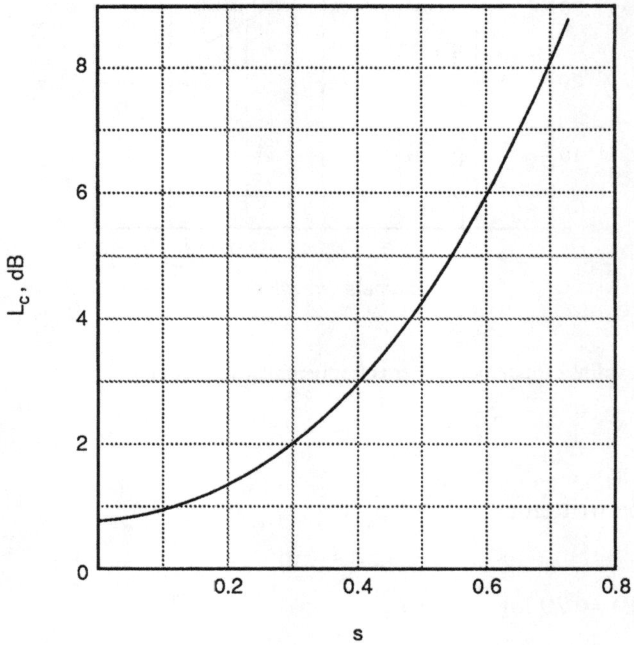

Figure 4.15 Gain-reduction factor for combined amplitude-taper loss and phase-error loss in conical horns. (Data are from Table 4.3.)

4.2.3 Phase Centers

A conical horn with a quadratic phase error across its aperture will have a phase center behind the aperture (within the horn), as was the case for rectangular horns. The locations were tabulated by Milligan [3] and are shown in Table 4.4. The distance from the aperture plane to the phase center (see Figure 4.9) divided by the wavefront radius of curvature is presented as a function of the normalized ray-path errors [defined in eq. (4.12)]. The data from Table 4.4 are presented graphically in Figure 4.16.

Table 4.4
Phase-Center Location of a Conical Horn (TE_{11} Mode) behind the Aperture as a Ratio of the Wavefront Radius of Curvature*

s	*H-plane* L_{ph}/R	*E-plane* L_{pe}/R	s	*H-plane* L_{ph}/R	*E-plane* L_{pe}/R
0.00	0.000	0.000	0.28	0.235	0.603
0.04	0.0046	0.012	0.32	0.310	0.782
0.08	0.018	0.048	0.36	0.397	
0.12	0.042	0.109	0.40	0.496	
0.16	0.075	0.194	0.44	0.604	
0.20	0.117	0.305	0.48	0.715	
0.24	0.171	0.441			

*After Milligan [3].

4.3 CORRUGATED HORNS

Corrugated horns [1, 3, 4, 10–12] have been developed during the last three decades. They have the desired properties of axially symmetric beams, low sidelobes, and low cross polarization; however, they have the disadvantages of being heavier, larger, and more expensive than smooth-walled horns. The two most popular types are small-flare-angle and large-flare-angle horns as shown in Figure 4.17.

4.3.1 Patterns

A universal set of patterns for small-flare horns is shown in Figure 4.18; the patterns are reasonably accurate for half-flare angles up to about 20 degrees. For small horns, the pattern amplitude should be multiplied by the obliquity factor [see eq. (4.3)]. For $2\pi a/\lambda < 6$, the flange (circular band of metal in the plane of and sur-

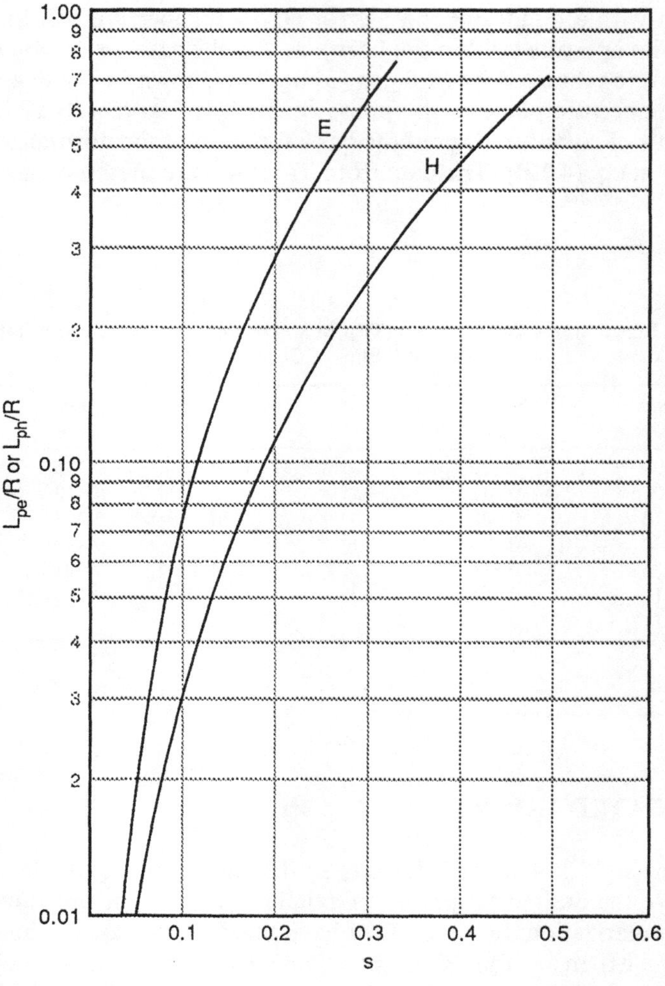

Figure 4.16 Phase-center location behind the aperture of a conical horn. (Data are from Table 4.4.)

rounding the aperture) at the aperture begins to affect the radiation pattern; and for $2\pi a/\lambda < 4$, the patterns are not valid.

A universal set of patterns for large-flare horns is shown in Figure 4.19; note that the abscissa is θ/θ_f where θ is the far-field angle and θ_f is the half-flare angle.

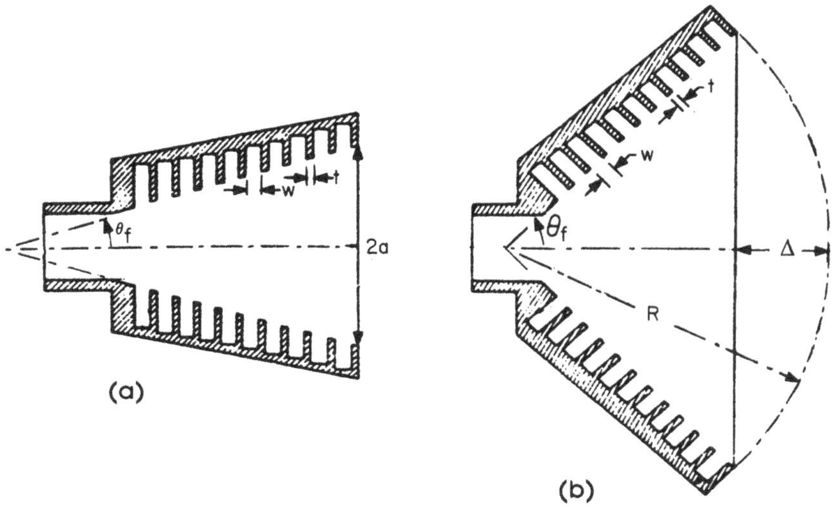

Figure 4.17 Conical corrugated horns: (a) small-flare-angle; (b) large-flare-angle. (©1978 IEEE, from Thomas [11].)

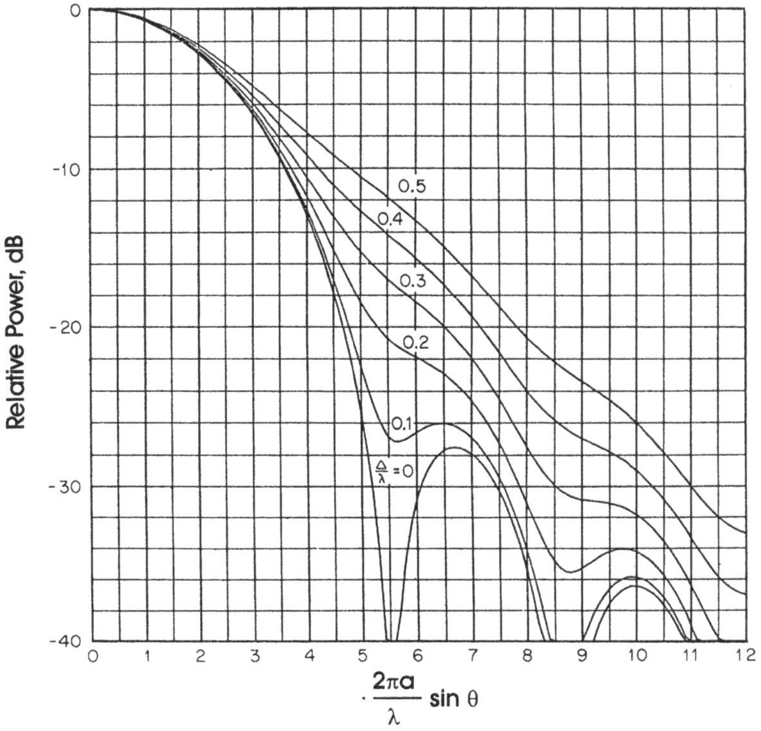

Figure 4.18 Universal patterns for small-flare-angle corrugated horns. (After Love [1], with permission of McGraw-Hill, Inc.)

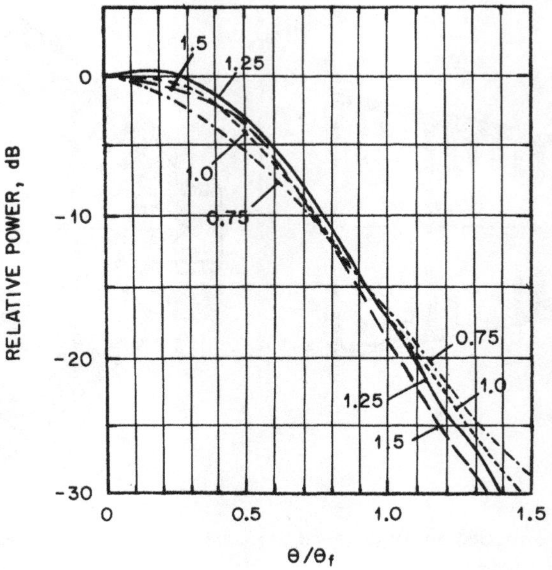

Figure 4.19 Universal patterns for large-flare-angle corrugated horns with Δ/λ as a parameter. (©1978 IEEE, from Thomas [11].)

Because θ/θ_f is independent of frequency, the patterns tend to be frequency insensitive. The patterns are applicable for values of θ_f up to about 70 degrees. A good rule of thumb is that the beam angle at the -12-dB level occurs at about $0.8\theta_f$, while the angle at the -15-dB level occurs at about $0.9\theta_f$ for any Δ between 0.75λ and 1.5λ.

In the design of most horns, we are concerned primarily with the shape of the main lobe. The designer can estimate the main beamshape quickly through the use of Figures 4.20 through 4.22. These figures can be used for both small-flare and large-flare horns.

4.3.2 Directivity and Gain

For most microwave horns, the losses are small, so the gain and directivity are essentially identical. At the higher frequencies, however, we may need to account for dissipative losses in the horn. In this section, we will assume that gain and directivity are equal.

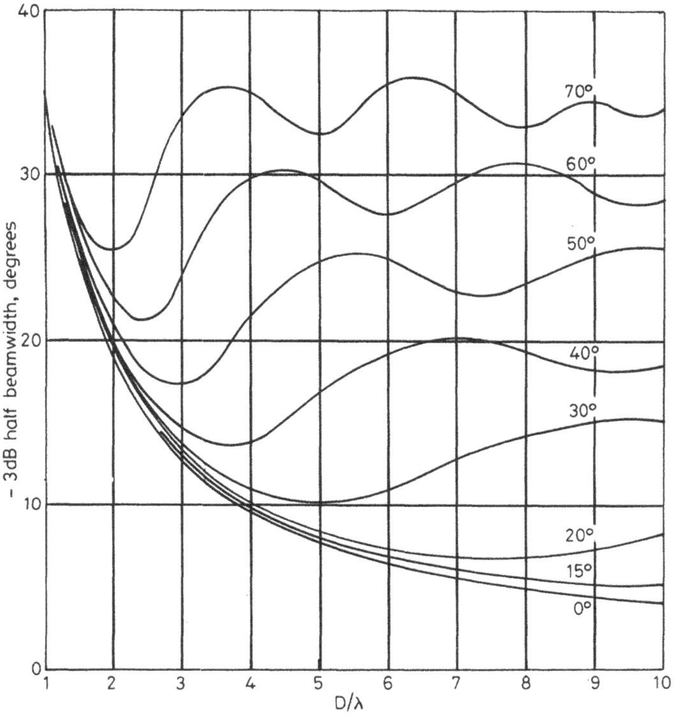

Figure 4.20 The −3-dB half-beamwidth of corrugated horns *versus* the normalized aperture diameter with the half-flare angle as a parameter. (From Clarricoats and Olver [12].)

The peak directivity has been computed (p. 135 [12]) for a wide range of aperture diameters and horn half-flare angles. The results are presented in Figure 4.23; the peak directivity of a uniformly illuminated circular aperture is also shown for comparison. This figure can be used for both small-flare and large-flare horns.

4.3.3 Phase Centers

For low cross-polar designs, the phase centers in the E- and H-planes are essentially at the same place; in this section, we will assume that they are coincident.

For open-ended corrugated waveguides (zero-flare angle), the phase center is essentially fixed at the center of the aperture of the waveguide. For large-flare-angle horns, the phase center is essentially fixed at the apex. For small-flare-angle horns,

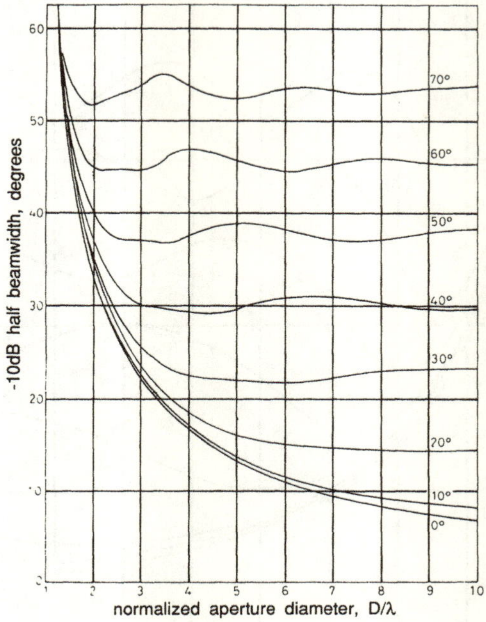

Figure 4.21 The −10-dB half-beamwidth of corrugated horns *versus* the normalized aperture diameter with the half-flare angle as a parameter. (From Clarricoats and Olver [12].)

the phase center will be somewhere between the apex and the aperture, and the position of the apparent phase center is a function of frequency, angle off boresight, and distance to the observation point.

Pertinent parameters are shown in Figure 4.24. The position of the phase center (observed in the far field) has been calculated [11] by minimizing the rms phase error (typically 3 degrees) between the −12-dB points of the main lobe; the results are shown in Figure 4.25.

4.3.4 Design of Corrugations

As indicated above, the co-polar radiation pattern, directivity, and phase-center location are determined essentially by the aperture diameter and the flare angle. The corrugations affect primarily the cross-polar radiation.

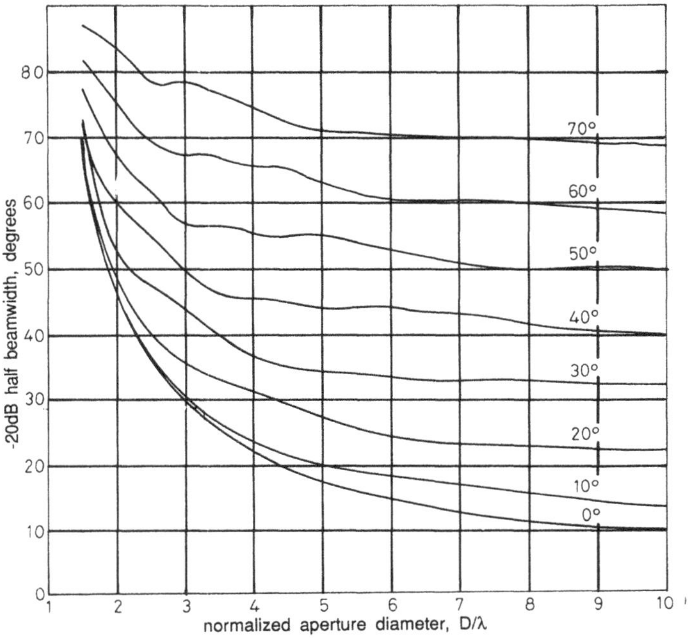

Figure 4.22 The −20-dB half-beamwidth of corrugated horns *versus* the normalized aperture diameter with the half-flare angle as a parameter. (From Clarricoats and Olver [12].)

In this discussion, let

d = slot depth
w = slot width
t = ridge width
p = period of corrugations ($p = w + t$)

Usually, the slots are cut normal to the axis of small-flare-angle horns and normal to the inside surface of large-flare-angle horns, as shown in Figure 4.17. The slot depth is taken to be the depth of the center of the slot.

Fortunately, the configuration of the corrugations is not critical; many configurations will yield satisfactory results. For minimum cross-polar radiation, the corrugation parameters should conform with those data in Figures 4.26 through 4.29. In designing a small-flare-angle horn, for example, we might arbitrarily select

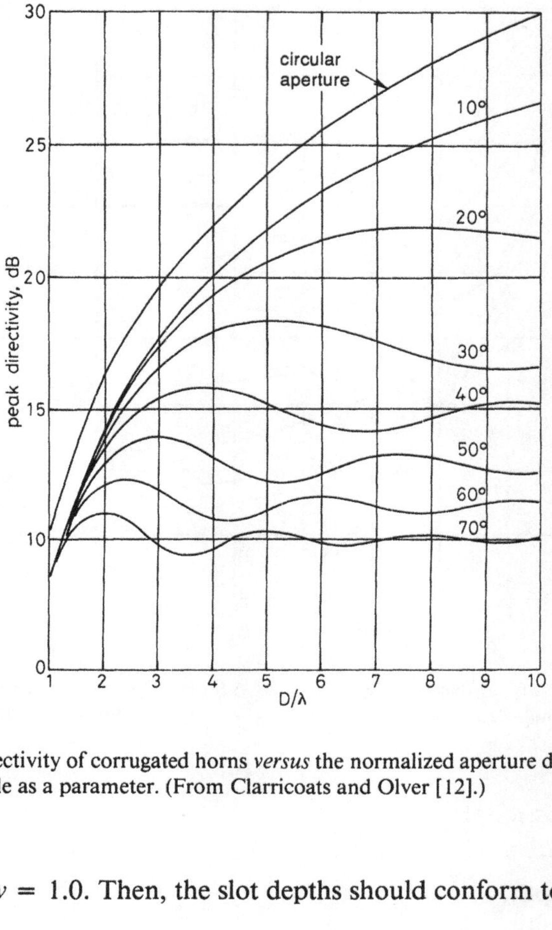

Figure 4.23 Peak directivity of corrugated horns *versus* the normalized aperture diameter with the half-flare angle as a parameter. (From Clarricoats and Olver [12].)

$w = 0.1\lambda$ and $t/w = 1.0$. Then, the slot depths should conform to the lower curve of Figure 4.26.

4.3.5 Impedance Matching

The throat region of a corrugated horn affects the impedance match. Usually, the horn is fed by a smooth-wall circular waveguide, and any mismatch at the throat can cause an unacceptable return loss [or voltage standing wave ratio (VSWR)]. Clarricoats and Olver (Section 5.5 [12]) presented a simple graphical method for designing the throat region, and their method is applicable to most corrugated horns.

Briefly, the horn is approximated by a series of constant diameter corrugated waveguides with each section containing one slot. The guide wavelength is high at

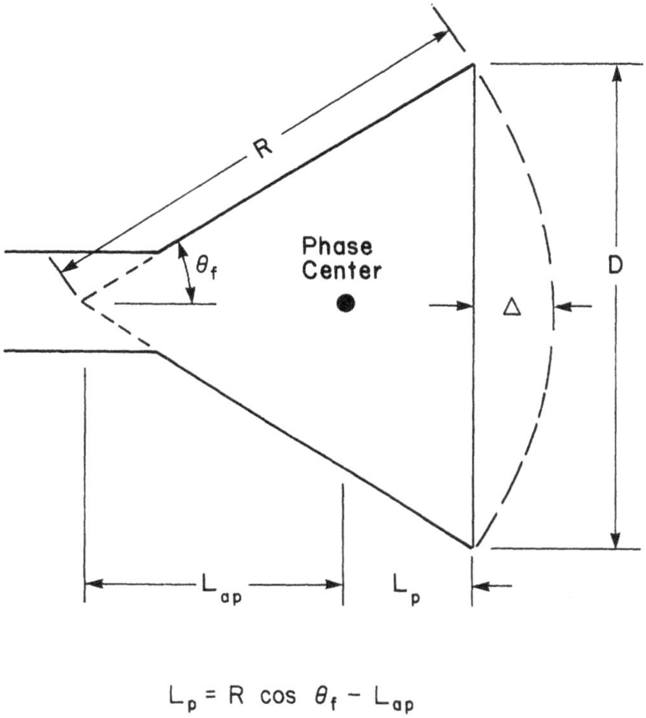

Figure 4.24 Parameters related to the location of the phase center of a corrugated horn; the corrugations are not shown.

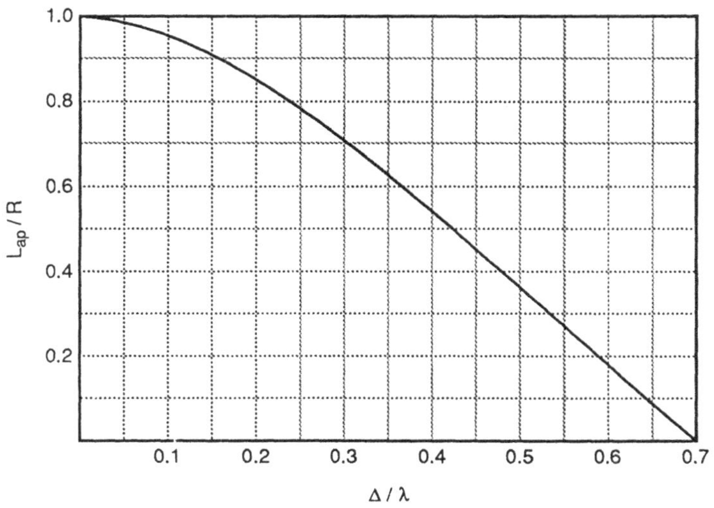

Figure 4.25 Distance of phase center from apex divided by slant length of a small-flare-angle corrugated horn. (©1978 IEEE, from Thomas [11].)

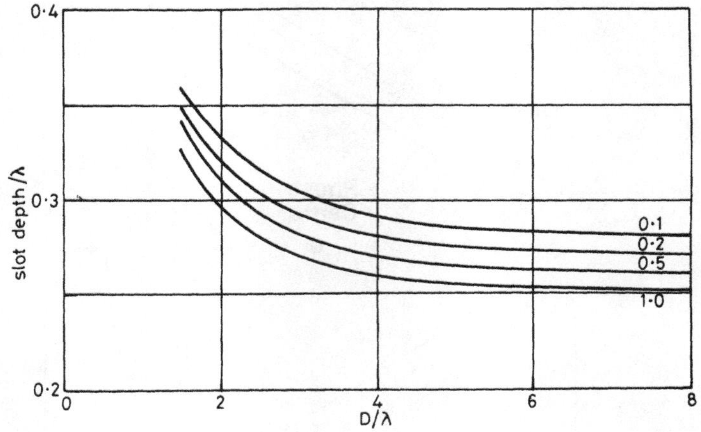

Figure 4.26 Slot depth for minimum cross polarization *versus* normalized aperture diameter. Slot width = 0.1λ. Parameter: ridge width-to-slot width ratio. (From Clarricoats and Olver [12].)

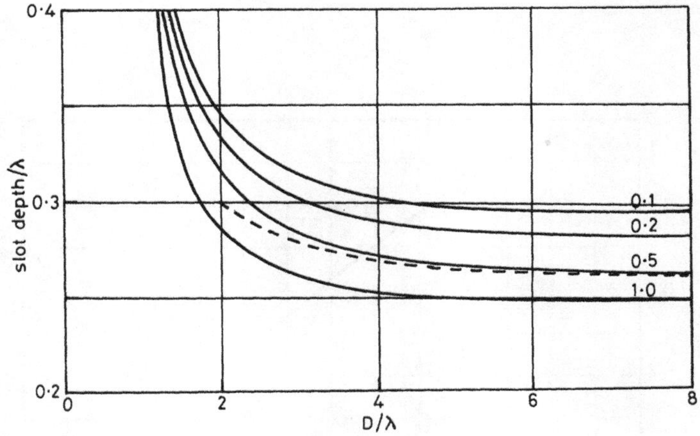

Figure 4.27 Slot depth for minimum cross polarization *versus* normalized aperture diameter. Slot width = 0.2λ. Parameter: ridge width-to-slot width ratio. Ignore the dashed curve. (From Clarricoats and Olver [12].)

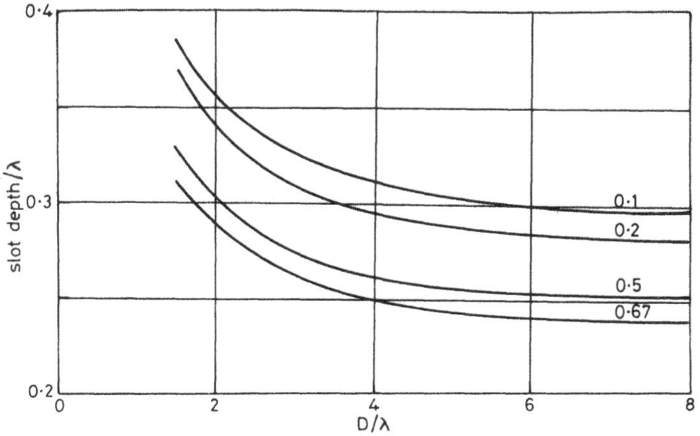

Figure 4.28 Slot depth for minimum cross polarization *versus* normalized aperture diameter. Slot width = 0.3λ. Parameter: ridge width-to-slot width ratio. (From Clarricoats and Olver [12].)

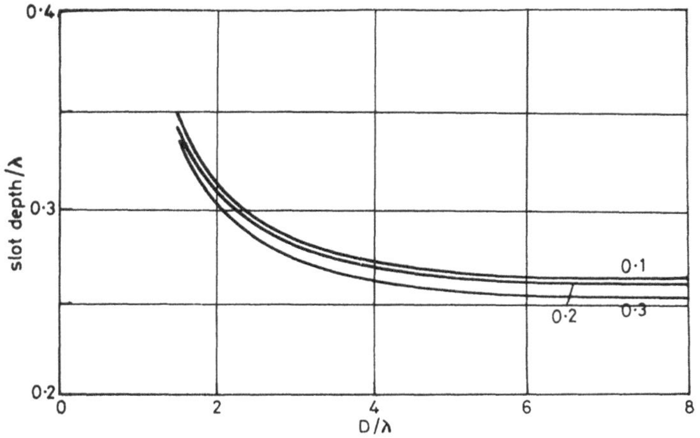

Figure 4.29 Slot depth for minimum cross polarization *versus* normalized aperture diameter. Ridge-width-to-slot-width ratio = 0.5. Parameter: slot width λ. (From Clarricoats and Olver [12].)

the throat junction, and it decreases toward the horn aperture. The change in guide wavelength with distance along the horn axis should be as low as possible.

The information needed to design the throat region is presented in Figure 4.30. The curves show the normalized guide wavelength *versus* the normalized slot depth for various normalized inner radii of the horn. These curves were calculated for a slot width of 0.1λ and a ridge width of 0.05λ; however, they can be used for most longitudinal slot geometries.

We used this graphical technique on a small-flare-angle horn having $D = 3.336\lambda$ and 25 slots with $w = t = 0.1\lambda$. The horn was fed with circular waveguides having a radius of 0.536λ. The slots were numbered from 1 to 25 starting at the aperture, and the depths were varied to conform with the lower curve of Figure 4.26. In the throat region, the depths of slots numbered 25 through 21 were adjusted to conform with the points indicated in Figure 4.30; note that their depths

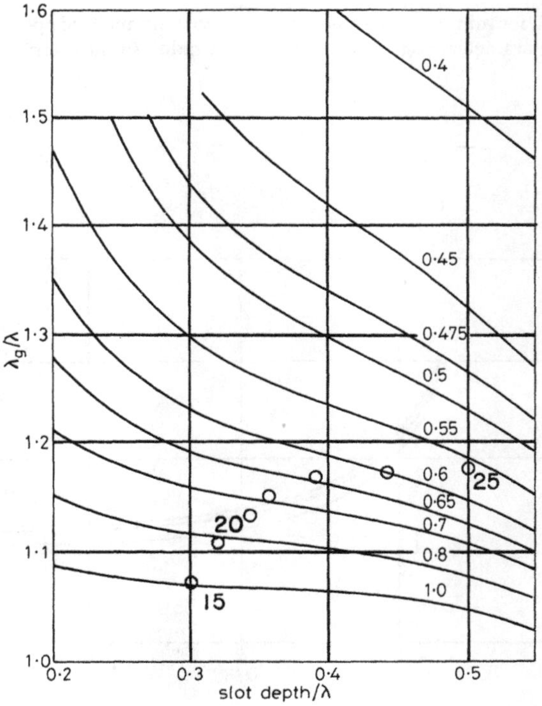

Figure 4.30 Normalized guide wavelength *versus* normalized slot depth for corrugated waveguide. Parameter: a/λ. Data for slots numbered 15 and 20–25 of an example design case are indicated with small circular points. (From Clarricoats and Olver [12].)

result in a gradual change of the normalized guide wavelength. Over the required 10% bandwidth, the maximum VSWR for the assembled rectangular-to-circular transducer and corrugated horn was 1.08.

REFERENCES

1. A.W. Love, "Horn Antennas," Chapter 15 in *Antenna Engineering Handbook,* 2nd Ed. (R.C. Johnson and H. Jasik, eds.), New York: McGraw-Hill, 1984.
2. J.R. Risser, "Waveguide and Horn Feeds," Chapter 10 in *Microwave Antenna Theory and Design* (S. Silver, ed.), MIT Radiation Laboratory Series, Vol. 12, New York: McGraw-Hill, 1949.
3. T.A. Milligan, *Modern Antenna Design,* New York: McGraw-Hill, 1985, Chap. 7.
4. A.W. Love, ed., *Electromagnetic Horn Antennas,* New York: IEEE Press, 1976.
5. E.V. Jull, "Finite Range Gain of Sectoral and Pyramidal Horns," *Electron. Lett.,* Vol. 6, October 15, 1970, pp. 680–681.
6. E.I. Muehldorf, "The Phase Center of Horn Antennas," *IEEE Trans. Antennas Propagat.,* Vol. AP-18, November 1970, pp. 753–760.
7. A.P. King, "The Radiation Characteristics of Conical Horn Antennas," *IRE Proc.,* Vol. 38, March 1950, pp. 249–251.
8. L.J. Chu, "Calculation of the Radiation Properties of Hollow Pipes and Horns," *J. Appl. Phys.* Vol. 11, 1940, pp. 603–610.
9. R.T. Compton and R.E. Collin, "Open Waveguides and Small Horns," Chapter 15 in *Antenna Theory, Part 1* (R.E. Collin and F.J. Zucker, eds.), New York: McGraw-Hill, 1969, p. 631.
10. B. MacA. Thomas, "Design of Wide-Band Corrugated Conical Horns for Cassegrain Antennas," *IEEE Trans. Antennas Propagat.,* Vol. AP-34, June 1986, pp. 750–757.
11. B. MacA. Thomas, "Design of Corrugated Conical Horns," *IEEE Trans. Antennas Propagat.,* Vol. AP-26, March 1978, pp. 367–372.
12. P.J.B. Clarricoats and A.D. Olver, *Corrugated Horns for Microwave Antennas,* London: Peter Peregrinus, 1984.

LIST OF SYMBOLS

A	H-plane dimension of rectangular-horn aperture
a	wide dimension of rectangular waveguide, radius of circular waveguide, or radius of aperture of corrugated horn
B	E-plane dimension of rectangular-horn aperture
b	narrow dimension of rectangular waveguide
D	aperture dimension (e.g., width, height, diameter)
d	corrugated slot depth
E_0	maximum magnitude of electric field
E_y	y-component of electric field
F_e	E-plane obliquity factor
F_h	H-plane obliquity factor
G	gain
G_0	area gain (with uniform phase)

j	imaginary symbol, $j = \sqrt{-1}$
k	$2\pi/\lambda$
K_e	gain reduction factor due to phase error in the E-plane
K_h	gain reduction factor due to phase error in the H-plane
L_{ap}	distance from horn apex to phase center
L_c	gain loss due to both amplitude taper and phase error in a circular conical horn, dB
L_e	gain loss due to phase error in the E-plane, dB
L_h	gain loss due to phase error in the H-plane, dB
L_p	distance of phase center behind aperture plane
L_{pe}	distance of E-plane phase center behind aperture plane
L_{ph}	distance of H-plane phase center behind aperture plane
p	period of corrugations
R	wavefront radius of curvature or slant length of horn from apex to aperture edge
R_e	wavefront radius of curvature in E-plane
R_h	wavefront radius of curvature in H-plane
s	normalized path error
s_e	E-plane normalized path error
s_h	H-plane normalized path error
t	corrugation ridge width
w	corrugation slot width
x, y, z	Cartesian coordinate variables
Δ	maximum path length difference for rays in horn aperture
Δ_e	maximum path length difference for rays in E-plane
Δ_h	maximum path length difference for rays in H-plane
η_a	aperture efficiency
θ	far-field pattern angle
θ_f	half-flare angle of horn
θ_{E10}	E-plane 10-dB beamwidth
θ_{H10}	H-plane 10-dB beamwidth
λ	wavelength
λ_g	guide wavelength

Chapter 5
REFLECTORS

Reflector antennas [1, 2] are the most common high-gain microwave antennas; this chapter will discuss some basic types.

5.1 PARABOLA

Most reflectors are based on the parabolic curve, which is illustrated in Figure 5.1. The equation for the parabola is

$$y^2 = 4Fx \tag{5.1}$$

The values of ρ and θ are

$$\rho = \sqrt{(F - x)^2 + y^2} \tag{5.2}$$

$$\theta = \arctan \frac{y}{F - x} \tag{5.3}$$

5.2 FRONT-FED PARABOLOID

A paraboloid is generated by rotating a parabola about its axis, and a paraboloidal reflector often is called a "dish." Such a reflector can be defined by the focal length F and the diameter D as illustrated in Figure 5.2.

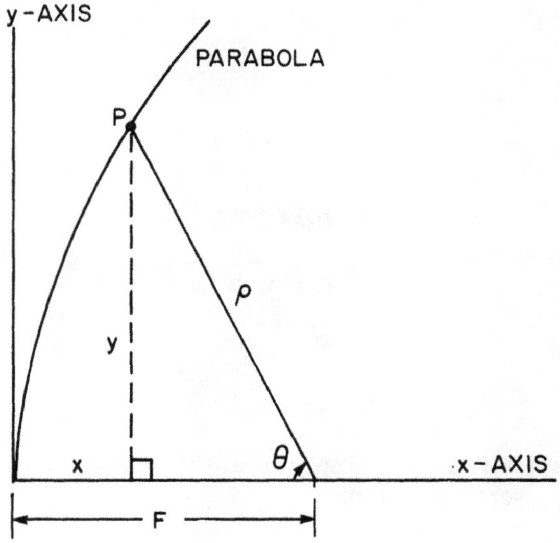

Figure 5.1 Parabola.

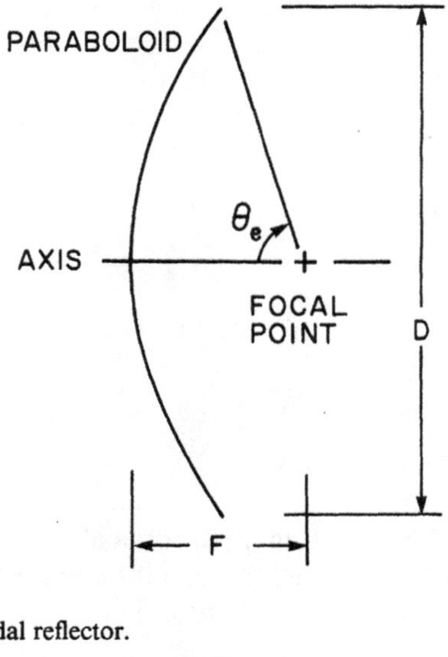

Figure 5.2 Front-fed paraboloidal reflector.

5.2.1 Subtended Angle

At the edge of the reflector, $\theta = \theta_e$ and $y = D/2$. The angle subtended by the reflector as viewed from the focal point is

$$2\theta_e = 2 \arctan \frac{D/2}{F - \dfrac{(D/2)^2}{4F}} = 2 \arctan \frac{8(F/D)}{16(F/D)^2 - 1} \tag{5.4}$$

The subtended angle versus F/D ratio is illustrated in Figure 5.3.

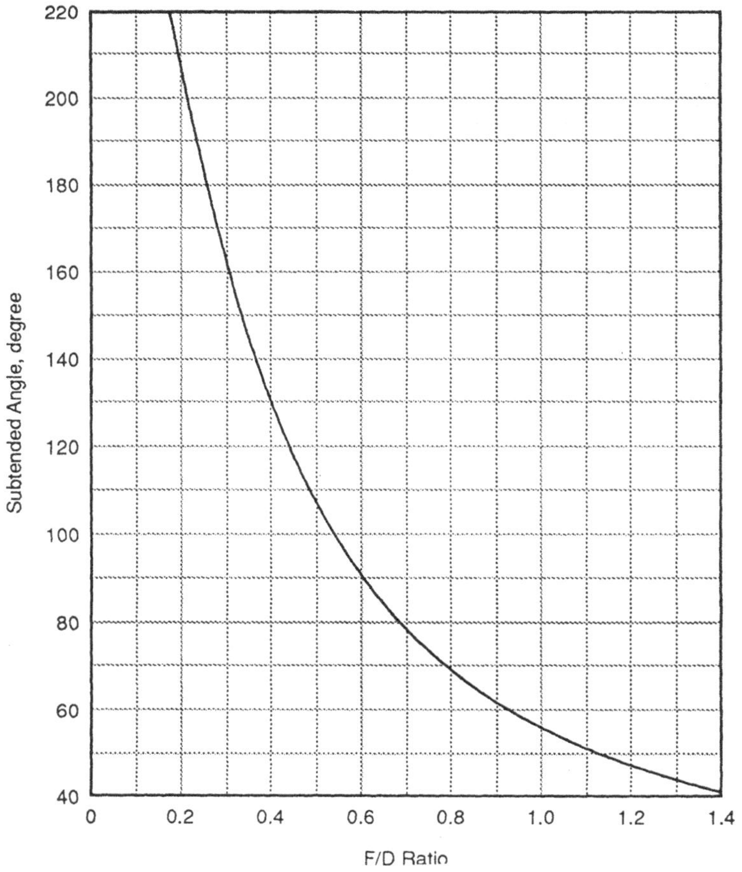

Figure 5.3 Subtended angle of paraboloidal reflector *versus* the F/D ratio.

5.2.2 Space Attenuation

As energy diverges from a point-source radiator, the power density decreases inversely with the square of the distance from the point source. Imagine an isotropic point source at the focal point of a paraboloidal reflector. The energy striking the vertex diverges a distance F, and the energy striking the reflector at any point away from the vertex diverges a distance greater than F. Thus, the reflected field is tapered from a reference value at the apex to a lesser value at the reflector's edge—even with an isotropic source as the feed. This phenomenon is called *space attenuation.* For a front-fed paraboloidal reflector, the space attenuation is

$$A_s = 20 \log \frac{\rho}{F} = 20 \log \sec^2\left(\frac{\theta}{2}\right) \tag{5.5}$$

The space attenuation *versus* the angle θ is illustrated in Figure 5.4. Note that if we are working with a parabolic-cylinder reflector and a line-source feed, the space

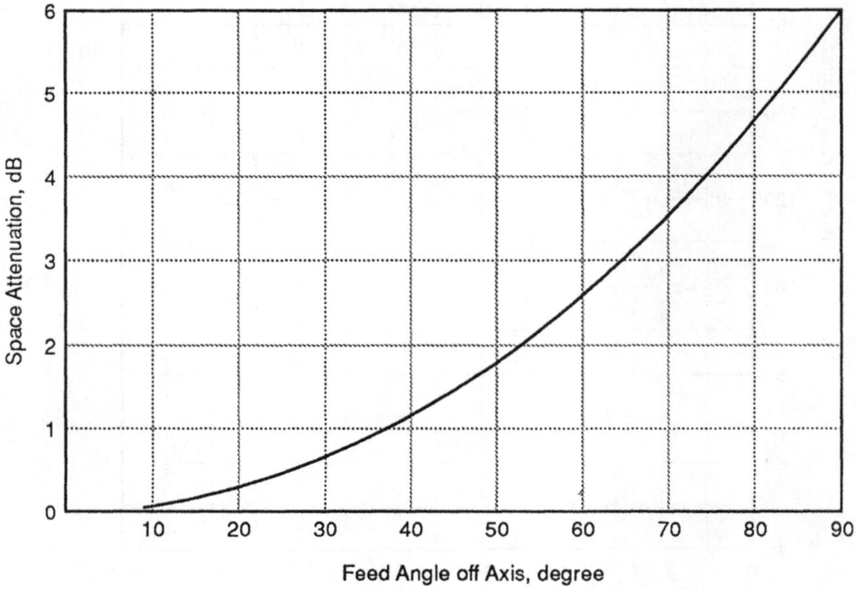

Figure 5.4 Space attenuation *versus* feed angle for a paraboloidal reflector.

attenuation is half of that indicated in eq. (5.5) and Figure 5.4 because the field is spreading only in one dimension.

The total field taper across the aperture of the reflector is the sum of two tapers: (1) the feed radiation pattern and (2) space attenuation.

5.2.3 Condon Lobes

The maximum cross-polarized lobes (usually called *Condon lobes*) from a front-fed paraboloidal reflector are in the ±45-degree planes as illustrated by Jones [3] in Figure 5.5. The peaks of the cross-polarized lobes occur approximately at the same angles as the first minima of the co-polarized patterns.

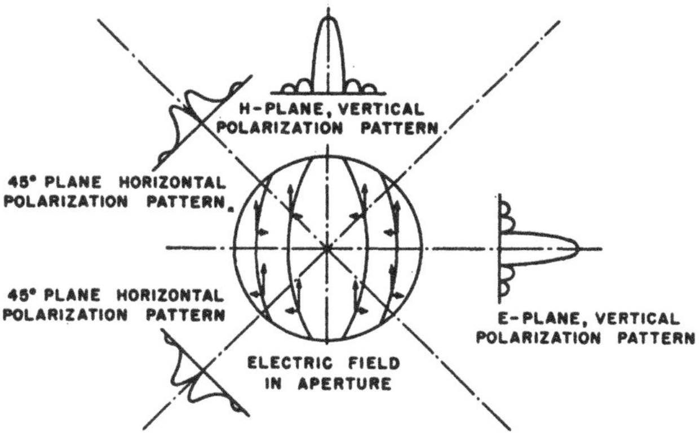

Figure 5.5 Electric field in the paraboloidal-reflector aperture and the resulting far-zone radiation patterns when the paraboloid is excited by a vertically oriented electric dipole. (©1954 IEEE, from Jones [3].)

The cross-polarized radiation depends primarily on the F/D ratio of the paraboloidal reflector and somewhat on the feed design. Typical peak levels of Condon lobes are illustrated in Figure 5.6.

5.2.4 Aperture Blockage

With a front-fed paraboloidal reflector, the feed and its support structure block portions of the aperture (also see the Appendix). The blockage can be assumed to be a field having an amplitude equal to the blocked field but out of phase by 180 degrees.

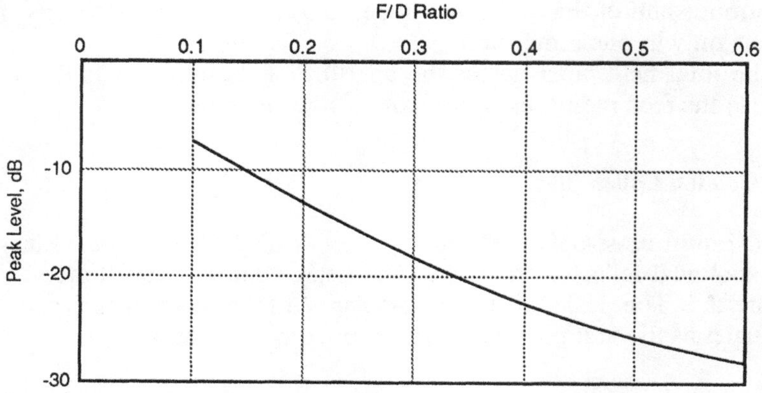

Figure 5.6 Approximate peak levels of the Condon (cross-polarized) lobes *versus* the *F/D* ratio when the paraboloid is excited by a short electric dipole. (Data from Jones [3].)

The net effects are a decrease in the main-lobe gain and an increase in the first sidelobe level [4].

Bodnar [5] presented graphical data for circular apertures having a circular blockage at the center; these data are shown in the Appendix. Gray [6] discussed the effects of strut blockage.

5.2.5 Beam-Deviation Factor

As the feed is moved off-axis, the main beam moves off-axis in the opposite direction as illustrated in Figure 5.7. The angular movement of the main beam is proportional to the angular movement of the feed [7] for small displacements

$$\theta_b = k_{bd}\theta_f \tag{5.6}$$

where k_{bd} is the beam-deviation factor.

This phenomenon was discussed briefly by Lo [8], and he presented the data in Figure 5.8. The beam-deviation factor is primarily a function of the *F/D* ratio and secondarily a function of the aperture illumination function.

5.3 OFFSET-FED PARABOLOID

The basic geometry of offset-fed paraboloidal-reflector antennas is illustrated in Figure 5.9; with the indicated orientation, the plane of symmetry is in elevation and

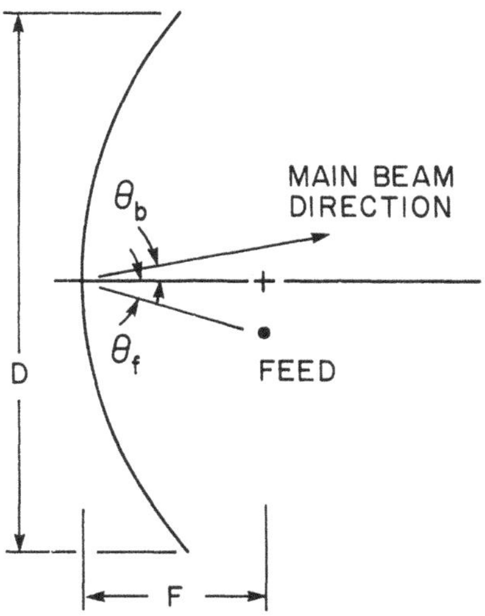

Figure 5.7 Sketch showing feed angle and main-beam angle.

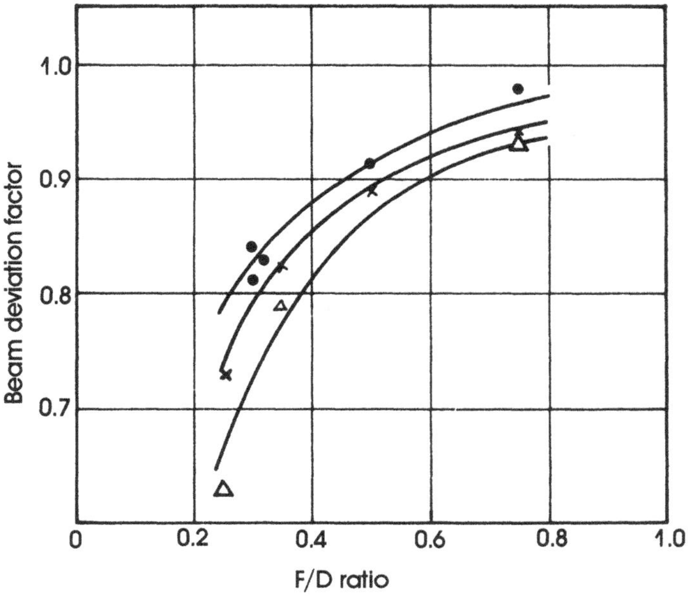

Figure 5.8 Beam-deviation factor *versus* the *F*/*D* ratio. Experimental data: • Silver and Pao [9], x 20-dB taper, and Δ 10-dB taper Kelleher and Coleman [10],—computed. (©1960 IEEE, from Lo [8].)

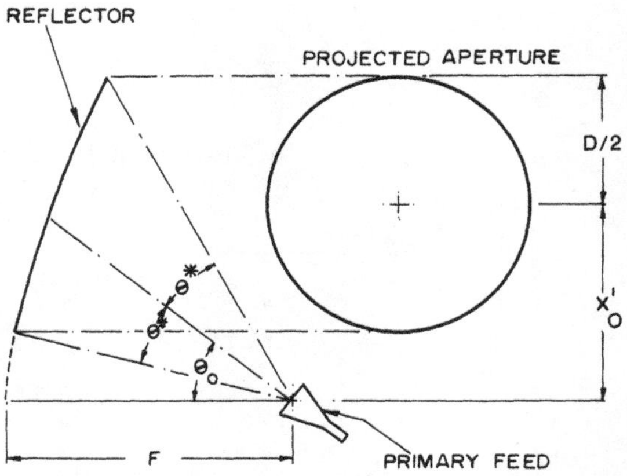

Figure 5.9 Basic geometry of offset-fed paraboloidal-reflector antenna. (From Rudge [11].)

the plane of asymmetry is in azimuth. With the angle θ^* held constant, a rotation about the feed axis generates a circular cone. This cone intersects the paraboloid with an elliptical contour, and this contour projects into a circular aperture. Such antennas have the advantages of little or no aperture blockage and reduced feed-reflector interaction. The disadvantages include larger and more expensive structures. An excellent discussion of these antennas was presented by Rudge [11].

5.3.1 Linear Polarization

A linearly polarized feed will produce a pair of cross-polar lobes in the asymmetric plane. The approximate positions of the peaks of the cross-polar lobes are at the −6-dB contour of the main co-polar beam. Some general sidelobe trends for fully offset-fed antennas are illustrated in Figure 5.10. Note that the larger offset angles result in higher levels of both co-polar and cross-polar sidelobes.

The depolarization properties of offset reflector antennas were studied by Chu and Turrin [12]. The maximum cross polarization for linearly polarized feed excitations is shown in Figure 5.11. A feed pattern with a 10-dB taper was used in their calculations for this figure; however, their calculations also showed that increasing the feed taper to 20 dB decreases the maximum cross-polar radiation by only about 1 dB.

5.3.2 Circular Polarization

When an offset-fed paraboloidal reflector is illuminated by a pure circularly polarized feed, no cross-polar radiation is generated, but the main beam is displaced by

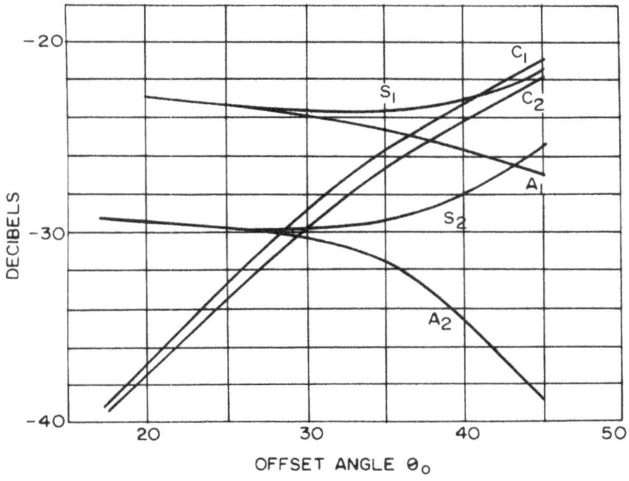

Figure 5.10 Peak co-polar sidelobe levels in planes of symmetry (S) and asymmetry (A) for offset reflectors with $\theta_0 = \theta^* + 5$ degrees fed by uniformly illuminated circular aperture feeds producing -10-dB (subscript 1) and -20-dB (subscript 2) illumination tapers at the reflector edge. Peak cross-polar levels (C) in the plane of asymmetry are also indicated. (From Rudge [11].)

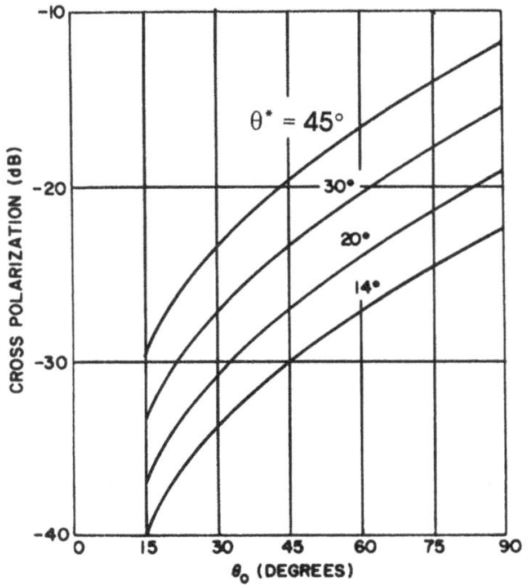

Figure 5.11 Maximum cross polarization of linearly polarized excitation. (©1973 IEEE, from Chu and Turrin [12].)

an angle $\Delta\beta$ from the plane of symmetry. The direction of displacement (as viewed from the reflector looking down the beam toward the far field) is toward the right for a far-field beam having left-handed circular polarization and toward the left for a far-field beam having right-handed circular polarization. The magnitude of the displacement is given in Figure 5.12.

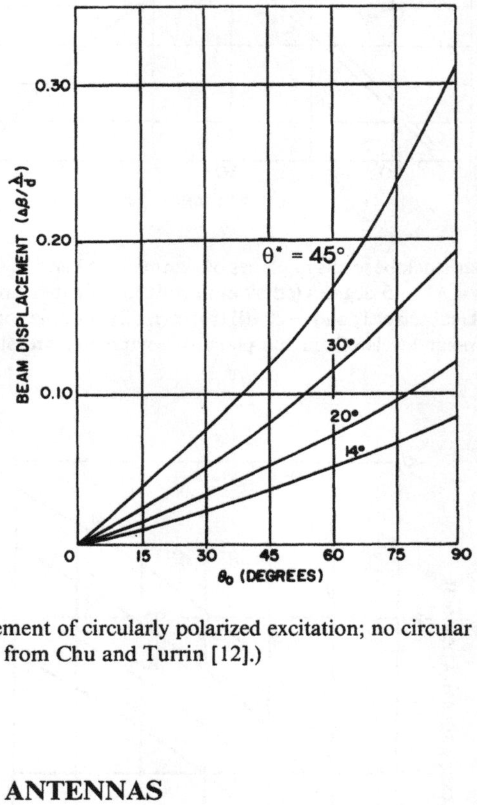

Figure 5.12 Beam displacement of circularly polarized excitation; no circular cross polarization exists. (©1973 IEEE, from Chu and Turrin [12].)

5.4 CASSEGRAIN ANTENNAS

Another popular high-gain antenna is the Cassegrain [2, 13, 14] which is a dual-reflector system as illustrated in Figure 5.13. This antenna is a fairly compact structure, which is very useful for many high-gain applications.

In the classical case, the main reflector is a paraboloid and the subreflector is a hyperboloid. For the best aperture efficiency, the blockage of the feed should be equal to the blockage of the subreflector as illustrated in Figure 5.14.

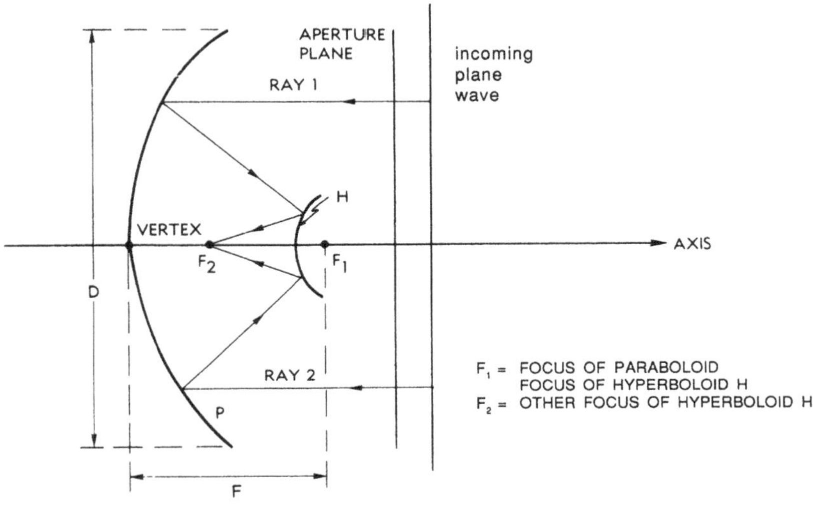

Figure 5.13 Geometry of a Cassegrain antenna. (From Kelleher and Hyde [2], with permission of McGraw-Hill, Inc.)

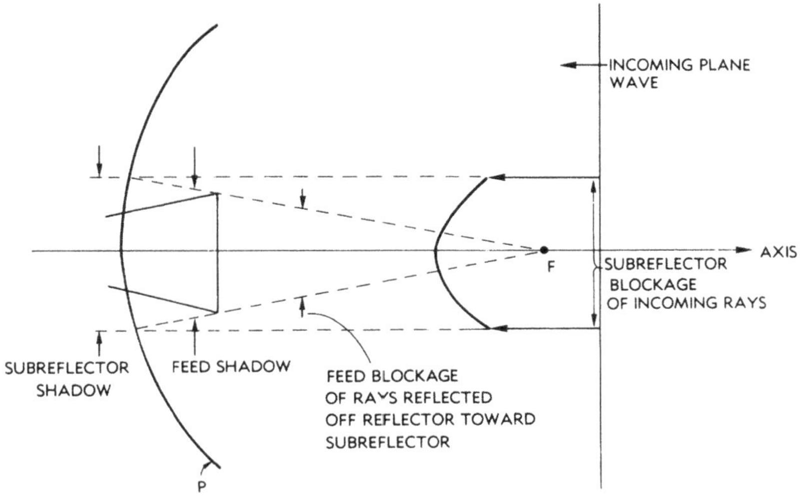

Figure 5.14 Blockage by feed and subreflector. (From Kelleher and Hyde [2], with permission of McGraw-Hill, Inc.)

REFERENCES

1. A.W. Love, ed., *Reflector Antennas,* New York: IEEE Press, 1978.
2. K.S. Kelleher and G. Hyde, "Reflector Antennas," Chapter 17 in *Antenna Engineering Handbook,* 2nd Ed. (R.C. Johnson and H. Jasik, eds.), New York: McGraw-Hill, 1984.
3. E.M.T. Jones, "Paraboloid Reflector and Hyperboloid Lens Antennas," *IEEE Trans. Antennas Propagat.,* July 1954, pp. 119–127.
4. S. Silver, *Microwave Antenna Theory and Design,* New York: McGraw-Hill, 1949, pp. 190–192.
5. D.G. Bodnar, "Materials and Design Data," Chapter 46 in *Antenna Engineering Handbook,* 2nd Ed. (R.C. Johnson and H. Jasik, eds.), New York: McGraw-Hill, 1984.
6. C.L. Gray, "Estimating the Effect of Feed Support Member Blocking on Antenna Gain and Sidelobe Level," *Microwave J.,* March 1964, pp. 88–91.
7. S. Silver, *Microwave Antenna Theory and Design,* New York: McGraw-Hill, 1949, p. 488.
8. Y.T. Lo, "On the Beam Deviation Factor of a Parabolic Reflector," *IEEE Trans. Antennas Propagat.,* May 1960, pp. 347–349.
9. S. Silver and C.S. Pao, "Paraboloid Antenna Characteristics as a Function of Feed Tilt," MIT Radiation Laboratory Report 479, 1944.
10. K.S. Kelleher and H.P. Coleman, "Off-Axis Characteristics of the Paraboloidal Reflector," NRL Report 4088, December 1952.
11. A.W. Rudge, "Offset Parabolic Reflector Antennas," Section 3.3 in *The Handbook of Antenna Design* (A.W. Rudge, K. Milne, A.D. Olver, and P. Knight, eds.), Vol. 1, London: Peter Peregrinus, 1982.
12. T.S. Chu and R.H. Turrin, "Depolarization Properties of Offset Reflector Antennas," *IEEE Trans. Antennas Propagat.,* May 1973, pp. 339–345.
13. P.W. Hannan, "Microwave Antennas Derived from the Cassegrain Telescope," *IEEE Trans. Antennas Propagat.,* March 1961, pp. 140–153.
14. P.A. Jensen, "Cassegrain Systems," Section 3.2 in *The Handbook of Antenna Design* (A.W. Rudge, K. Milne, A.D. Olver, and P. Knight, eds.), Vol. 1, London: Peter Peregrinus, 1982.

LIST OF SYMBOLS

A_s	space attenuation
B	blockage ratio for circular aperture and centrally located circular blockage
D	aperture diameter
D_b	blockage diameter
F	focal length
k_{bd}	beam-deviation factor
x,y	Cartesian coordinate variables
$\Delta\beta$	displacement (squint) angle of main lobe from a circularly polarized offset-fed paraboloidal reflector
θ	angle subtended from the vertex to a general point on a parabola as viewed from the focal point
θ_b	main-beam angle relative to the reflector axis
θ_e	angle subtended from the vertex to the edge of the reflector as viewed from the focal point

θ_f	feed angle relative to the reflector axis
θ_0	offset (tilt) angle of the feed in an offset-fed paraboloidal reflector
θ^*	half-angle subtended by an offset-fed paraboloidal reflector as viewed from the focal point
ρ	distance from the focal point to a general point on a paraboloid

Chapter 6
RADOMES AND LENSES

6.1 INTRODUCTION

Radomes [1–4] are dielectric structures that are used to protect antennas from the environment (wind, rain, dust, insects, *et cetera*), and they have a wide variety of shapes and wall structures. For some applications, the design of the radome is more complex than the design of the antenna to be protected. This chapter will address only simple radomes; however, they are applicable to many different situations.

When an electromagnetic wave is incident on an interface between two homogeneous dielectric regions, part of the energy is reflected and part of it is refracted as depicted in Figure 6.1. Usually, we assume that the interface is planar in the immediate vicinity of each ray under consideration. The plane that contains the incident ray and which is perpendicular to the interface is called the *plane of incidence.* The angle between the incident ray and the normal to the interface is called the *angle of incidence* (θ in Figure 6.1). If the electric field is in the plane of incidence, the polarization is said to be parallel, and if the electric field is normal to the plane of incidence, the polarization is said to be perpendicular.

The reflection coefficient at an interface is dependent on the polarization—especially at large angles of incidence. This phenomenon is illustrated with the example in Figure 6.2. The angle for which the reflection coefficient for parallel polarization becomes zero is called the *Brewster angle;* it occurs when

$$\theta_{\mathrm{B}} = \arctan\left(\frac{\epsilon_b}{\epsilon_a}\right)^{1/2} \tag{6.1}$$

where ϵ_a and ϵ_b are the permittivities of the materials on the incident and the refracted sides of the interface plane, respectively, and where $\epsilon_a < \epsilon_b$.

The two most important characteristics of radome materials are the dielectric constant ($\kappa = \epsilon/\epsilon_0$) and the loss tangent ($\tan\delta$). Some popular radome materials are listed in Tables 6.1 and 6.2. Note that they all have small loss tangents.

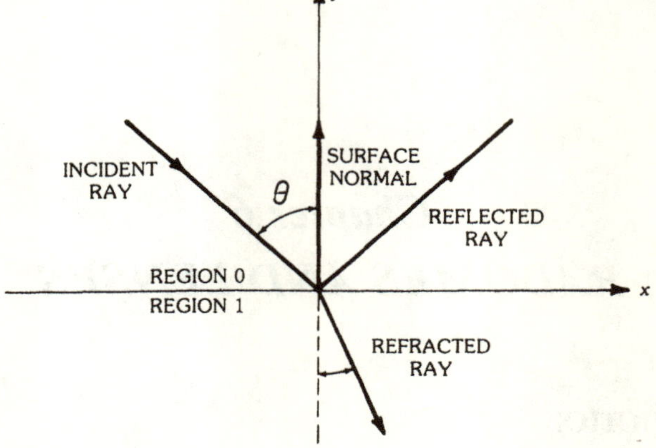

Figure 6.1 Reflection and refraction at a plane interface between two homogeneous dielectric regions. (After Tricoles [3].)

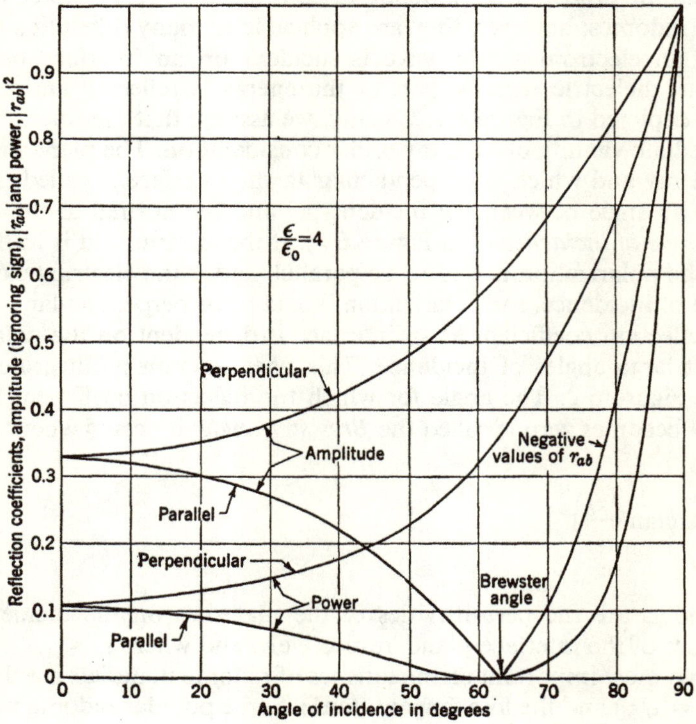

Figure 6.2 Reflection at oblique incidence.

Table 6.1
Organic Radome Materials (8.5 GHz)*

Material	*Dielectric Constant*	*Loss Tangent*
Thermal plastics		
Lexan	2.86	0.006
Teflon (PTFE)	2.10	0.0005
Noryl	2.58	0.005
Kydox	3.44	0.008
Laminates		
Epoxy-E glass cloth	4.40	0.016
Polyester-E glass cloth	4.10	0.015
Polyester-quartz cloth	3.70	0.007
Polybutadiene	3.83	0.015
Fiberglass laminate polybenzimidazole resin	4.9	0.008
Quartz-reinforced polyimide	3.2	0.008
Duroid 5650 (loaded PTFE)	2.65	0.003

*From Huddleston and Bassett [1].

Table 6.2
Ceramic Radome Materials (8.5 GHz)*

Material	*Density (g/cm^3)*	*Dielectric Constant*	*Loss Tangent*
Aluminum oxide	3.32	7.85	0.0005
Alumina, hot-pressed	3.84	10.0	0.0005
Beryllium oxide	2.875	6.62	0.001
Boron nitride, hot-pressed	2.13	4.87	0.0005
Boron nitride, pyrolytic	2.14	5.12	0.0005
Magnesium aluminate (spinel)	3.57	8.26	0.0005
Magnesium aluminum silicate (cordierite ceramic)	2.44	4.75	0.002
Magnesium oxide	3.30	9.72	0.0005
Pyroceram 9606		5.58	0.0008
Rayceram 8		4.72	0.003
Silicon dioxide	2.20	3.82	0.0005
Silica-fiber composite (AS-3DX)	1.63	2.90	0.004
Slip-cast fused silica	1.93	3.33	0.001
Silicon nitride	2.45	5.50	0.003

*From Huddleston and Bassett [1].

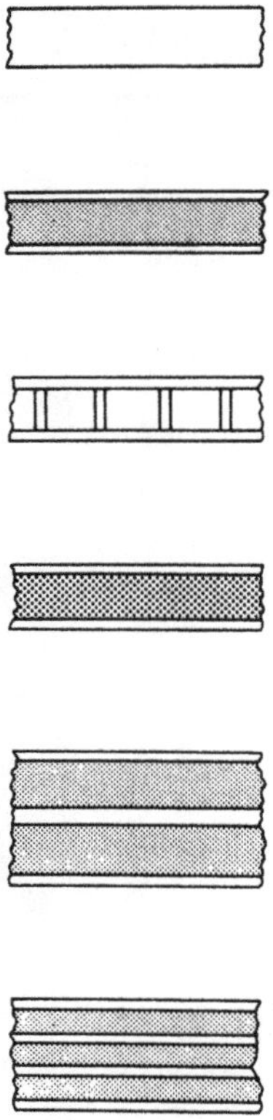

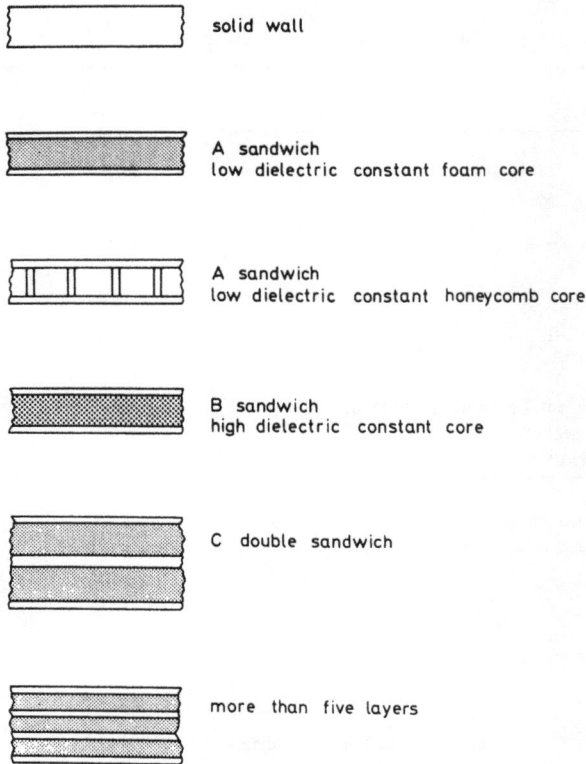

Figure 6.3 Radome wall constructions. (From Cary [2].)

Some basic radome wall structures are depicted in Figure 6.3. In this chapter, we will discuss only the solid wall and the A-sandwich.

6.2 SOLID-WALL RADOMES

A normal-incidence radome is one in which the radiation from the antenna falls almost normally on the radome wall. If the angle of incidence is less than about 30 degrees, the percentage of the incident power reflected by the radome is nearly the same as that reflected for normal incidence. Thus, most radomes are designed for normal incidence.

Figure 6.4 [5] presents the amplitude reflection and the power transmission coefficients versus wall thickness for various loss tangents and for a wall dielectric

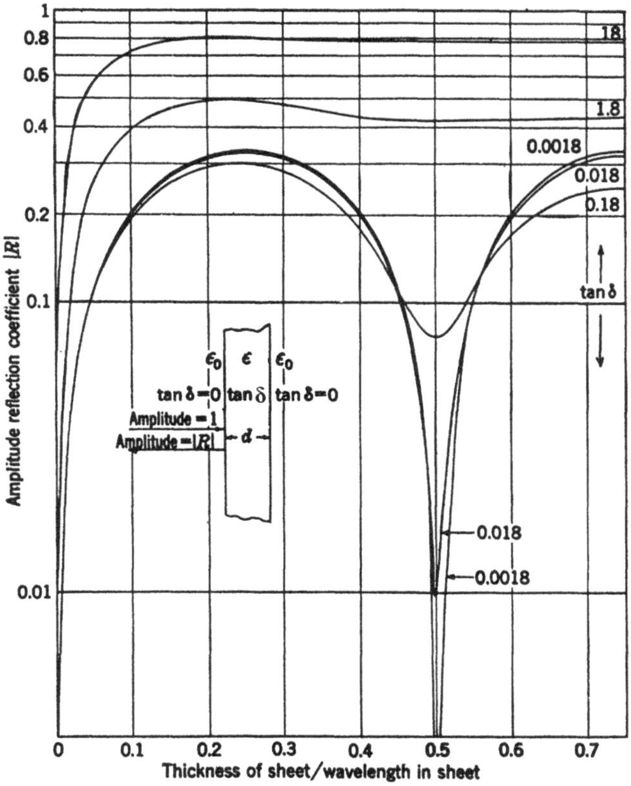

Figure 6.4 Amplitude reflection and power transmission coefficients for a plane lossy wall having a dielectric constant of $\kappa = 2$. (From Leaderman [5].)

constant of $\kappa = 2$. Figure 6.5 [5] presents similar data for a wall dielectric constant of $\kappa = 4$. The radome designer desires a small reflection coefficient and a large transmission coefficient; thus, the wall thickness should be less than about $\lambda/10$ (thin-wall radome) or about $\lambda/2$ (half-wavelength wall radome). Using these two figures, the designer should be able to select a material and wall thickness for satisfactory performance. Note that the thickness of the wall can be any integer multiple of $\lambda/2$, but it is rare to require a thickness greater than $\lambda/2$.

Other useful design data for thin radomes are presented in Figure 6.6. The designer can quickly choose a wall thickness (in free-space wavelengths) for 90 or 95% power transmission. Note that losses in the radome are neglected; however, they are usually not significant in thin radomes.

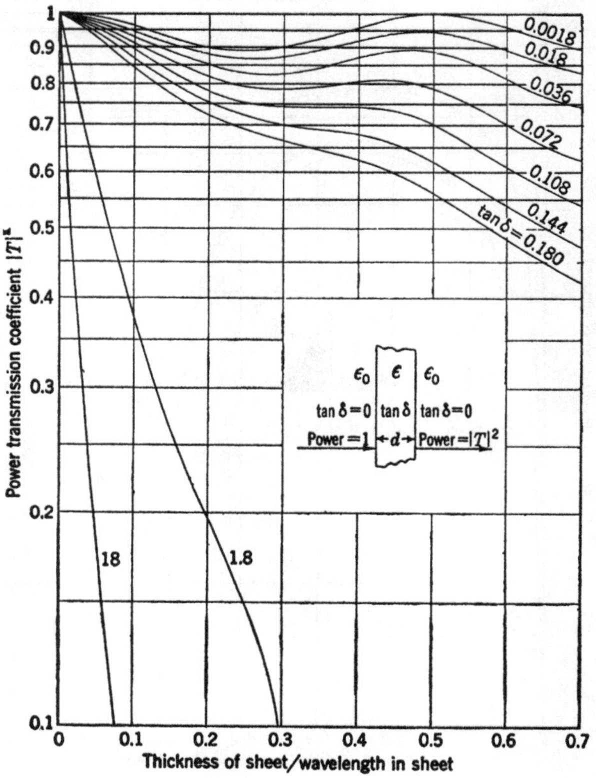

Figure 6.4 continued

If we wish to design a half-wavelength wall for an angle of incidence other than normal (i.e., $\theta \neq 0$), the wall thickness should be

$$d = \frac{\lambda_0}{2} \frac{1}{(\kappa - \sin^2\theta)^{1/2}} \tag{6.2}$$

The graphical data presented in Figures 6.4 through 6.6 allow us to design quickly a solid-wall radome for most applications. If we desire a more accurate design procedure, the power-reflection, power-transmission, and power-loss coefficients can be calculated as follows [6]:

$$|R|^2 = \frac{|r_{ab}|^2[(1 - A^2)^2 + 4A^2 \sin^2\phi]}{(1 - A^2|r_{ab}|^2)^2 + 4A^2|r_{ab}|^2 \sin^2(\phi + \chi)} \tag{6.3}$$

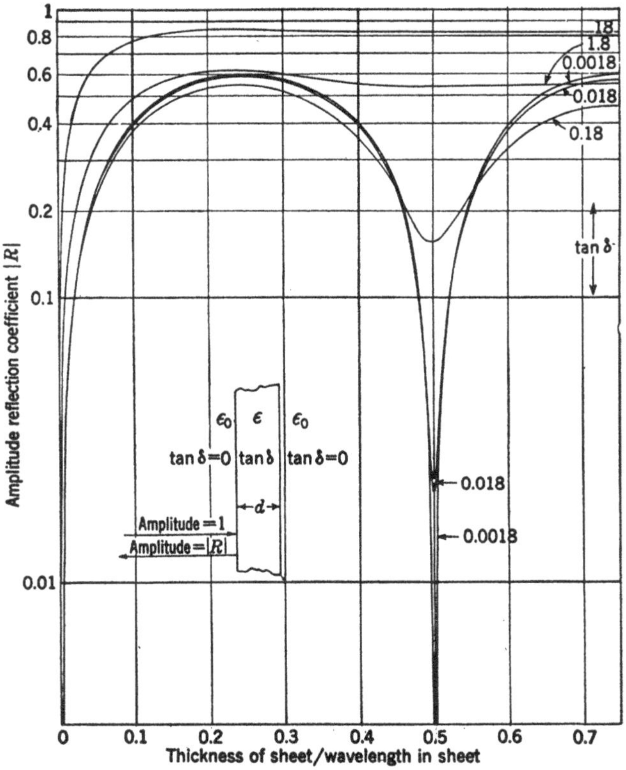

Figure 6.5 Amplitude reflection and power transmission coefficients for a plane lossy wall having a dielectric constant of $\kappa = 4$. (From Leaderman [5].)

$$|T|^2 = \frac{A^2[(1 - |r_{ab}|^2)^2 + 4|r_{ab}|^2 \sin^2\chi]}{(1 - A^2|r_{ab}|^2)^2 + 4A^2|r_{ab}|^2 \sin^2(\phi + \chi)} \tag{6.4}$$

where

$$A = \exp\left[-\frac{2\pi d}{\lambda_0}\frac{\kappa k}{(\kappa - \sin^2\theta)^{1/2}}\right] \tag{6.5}$$

$$\phi = \frac{2\pi d}{\lambda_0}(\kappa - \sin^2\theta)^{1/2} \tag{6.6}$$

$$k = \frac{\tan\delta}{2}\left(1 - \frac{\tan^2\delta}{4}\right) \tag{6.7}$$

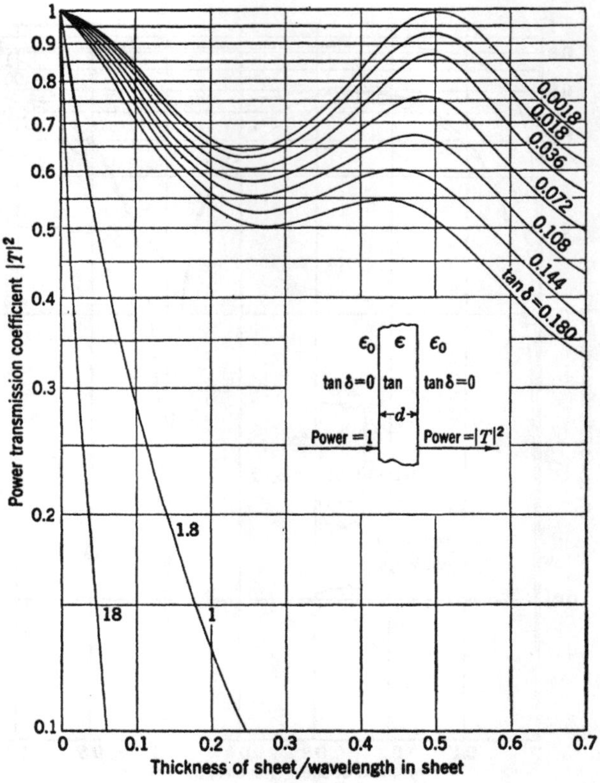

Figure 6.5 continued

$$\chi = \arctan\left(\frac{2\kappa^{1/2}k}{\kappa(1 - k^2) - 1}\right) \tag{6.8}$$

$$|r_{ab}| = \frac{\kappa^{1/2} - 1}{\kappa_{1/2} + 1} \tag{6.9}$$

From conservation of energy, the power-loss coefficient is

$$|L|^2 = 1 - |R|^2 - |T|^2 \tag{6.10}$$

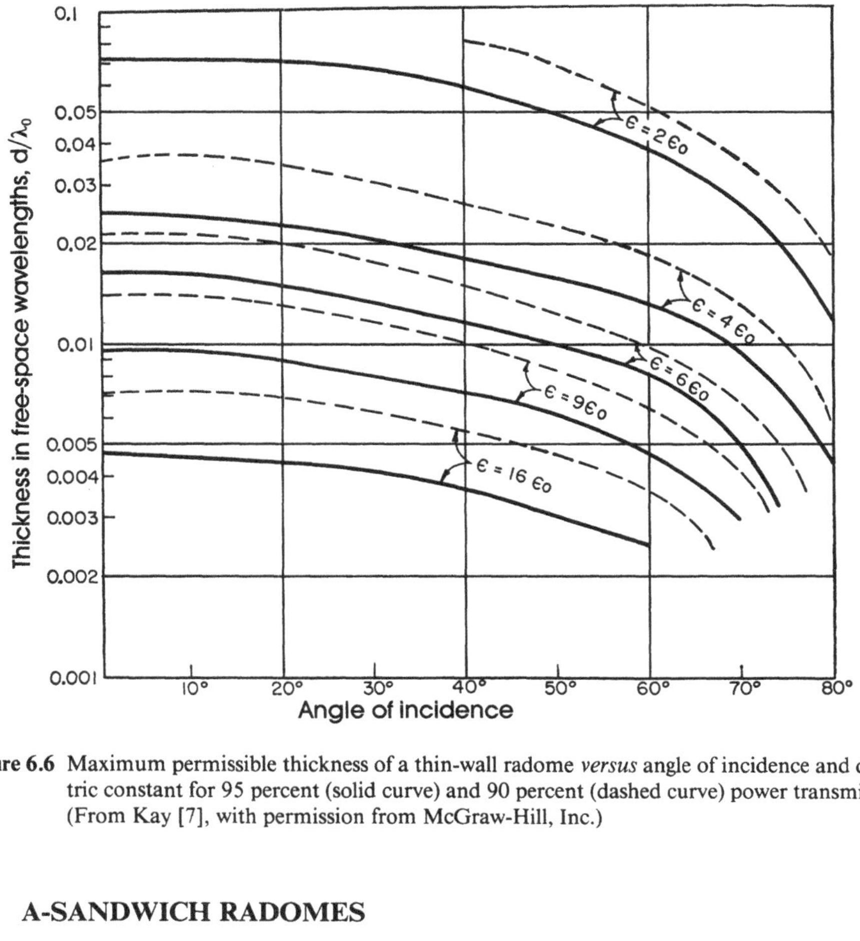

Figure 6.6 Maximum permissible thickness of a thin-wall radome *versus* angle of incidence and dielectric constant for 95 percent (solid curve) and 90 percent (dashed curve) power transmission. (From Kay [7], with permission from McGraw-Hill, Inc.)

6.3 A-SANDWICH RADOMES

An A-sandwich radome is a three-layer structure as depicted in Figure 6.3. It consists of a low dielectric-constant core with a protective skin on each side. If losses are neglected (a reasonable assumption for many radome materials), the reflection coefficient will be zero if the core thickness is [7]:

$$D = \frac{\lambda_0}{2\pi(\kappa_c - \sin^2\theta)^{1/2}} \times \left\{ n\pi - \arctan\left[\frac{2(\kappa_{se} - 1)(\kappa_{se}\kappa_{ce})^{1/2}\sin 2\phi}{(\kappa_{ce} - \kappa_{se})(1 + \kappa_{se}) + (\kappa_{se} - 1)(\kappa_{ce} + \kappa_{se})\cos 2\phi}\right]\right\} \tag{6.11}$$

where

$$\left.\begin{aligned} \kappa_{se} &= \frac{\kappa_s - \sin^2\theta}{\cos^2\theta} && \text{(perpendicular polarization)} \\ \kappa_{se} &= \frac{\kappa_s^2 \cos^2\theta}{\kappa_s - \sin^2\theta} && \text{(parallel polarization)} \end{aligned}\right\} \quad (6.12)$$

$$\left.\begin{aligned} \kappa_{ce} &= \frac{\kappa_c - \sin^2\theta}{\cos^2\theta} && \text{(perpendicular polarization)} \\ \kappa_{ce} &= \frac{\kappa_c^2 \cos^2\theta}{\kappa_c - \sin^2\theta} && \text{(parallel polarization)} \end{aligned}\right\} \quad (6.13)$$

$$\phi = \frac{2\pi\, d}{\lambda_0}(\kappa_s - \sin^2\theta)^{1/2} \quad (6.14)$$

and

n = an integer such that D is positive (usually 1)
d = skin thickness
D = core thickness
κ_c = dielectric constant of core
κ_s = dielectric constant of skin
θ = angle of incidence

These equations appear to be accurate for angles of incidence up to 75 degrees or greater.

6.4 LENSES

Lenses are collimating devices having up to four degrees of freedom (inner surface; outer surface; index of refraction (n); and, for constrained lenses, inner- *versus* outer-surface radiator positions). They have no aperture blockage by a feed, but have internal and surface-reflection losses, must be supported at the edges, and are relatively heavy and bulky. Many types of lenses have been developed [8–12], but only a few simple configurations will be discussed here.

The index of refraction can be written as

$$n = \frac{c}{v} = \left(\frac{\epsilon}{\epsilon_0}\right)^{1/2} = (\kappa)^{1/2} \quad (6.15)$$

$$r = \frac{(n-1)\,f}{n\cos\theta - 1}$$

$$y = \sqrt{(n^2-1)x^2 + 2(n-1)f\,x}$$

(a)

$$r = \frac{(n-1)f_1 - (K-1)\lambda_0}{n\cos\theta - 1}$$

$$y = \Big\{(n^2-1)x^2 + 2(n-1)f_1 x + 2(K-1)\lambda n x$$
$$+ 2f_1(K-1)\lambda + (K-1)^2 \lambda^2\Big\}^{1/2}$$

$$K = 1,2,3,4,\cdots$$

(b)

CURVED
SURFACE
SAME AS IN (a)

$$\Delta = \frac{\lambda}{n-1}$$

(c)

$$r_1 = f - d = \text{CONSTANT}$$

$$r_2 = \frac{(n-1)f}{n - \cos\theta_2}$$

$$y_1 = \sqrt{(f-d)^2 - (x_1+f)^2}$$

$$y_2 = \Big\{-2\Big(\frac{n-1}{n}\Big)f\,x_2 - \Big(\frac{n^2-1}{n^2}\Big)x_2^2\Big\}^{1/2}$$

(d)

SAME AS (f), WITH K = 1

(e)

$$x_2 = \left\{\frac{(K-1)\lambda + (n-1)d + f - \sqrt{f^2+y_1^2}}{n - \Big(1 - \frac{y_1^2}{n^2(f^2+y_1^2)}\Big)^{1/2}}\right\} \cdot \left[1 - \frac{y_1^2}{n^2(f^2+y_1^2)}\right]^{1/2}$$

$$y_2 = y_1\left[1 + \frac{x_2}{\big[n^2(f^2+y_1^2) - y_1^2\big]^{1/2}}\right], \quad K = 1,2,3,4,\cdots$$

(f)

for $n > 1$ media.

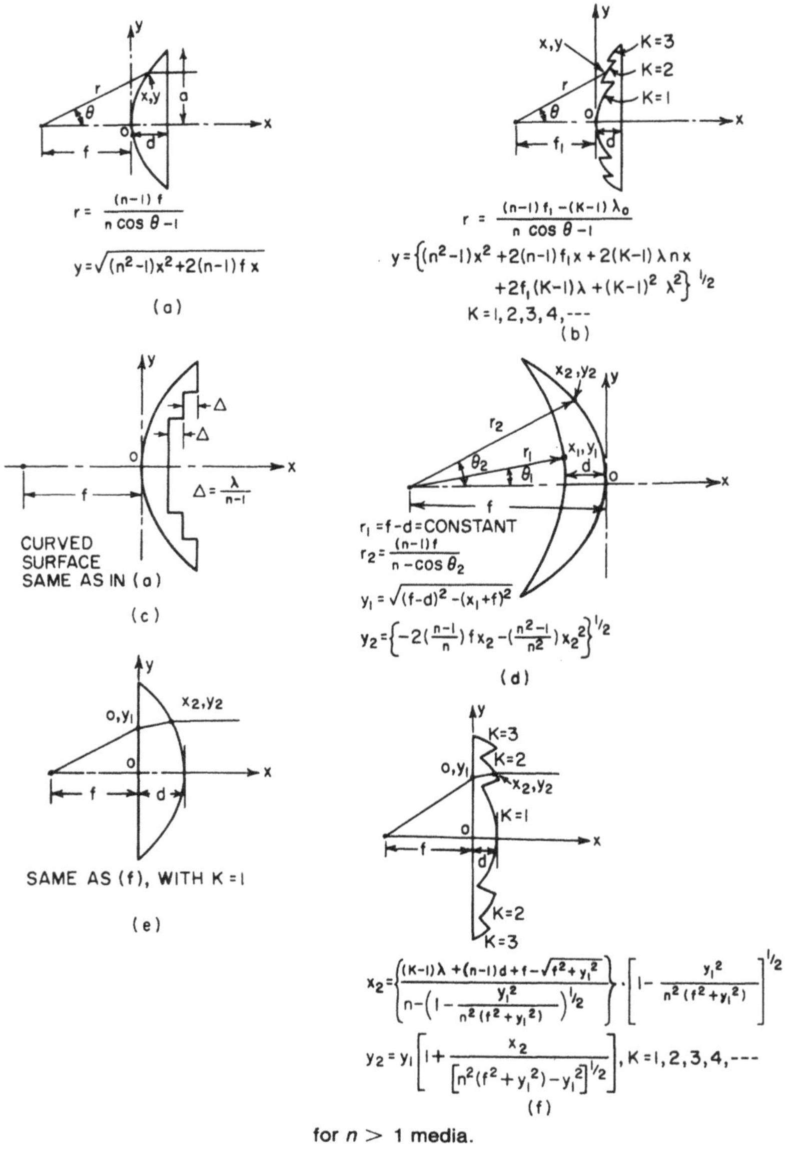

Figure 6.7 Design of lens surfaces for $n > 1$ media. (From Peeler [8], with permission of McGraw-Hill, Inc.)

where c is wave velocity in free space, v is wave velocity in the dielectric, ϵ is permittivity of the dielectric, ϵ_0 is permittivity of free space, and κ is the dielectric constant. Common shapes for $n > 1$ media lenses are shown in Figure 6.7. All of the illustrated lenses have only one degree of freedom because either they refract at only one surface or one surface is fixed during the design.

The lens designs of Figures 6.7(a) through (d) have one surface that is coincident with a wavefront. Reflections from this surface will arrive back at the feed in phase, and thus may produce an unacceptable VSWR in the feed line. One possible technique for reducing the reflections is to use a quarter-wavelength matching transformer at the lens surface. The transformer can consist of a quarter-wavelength thick coating of material having an index of refraction equal to the square root of n, where n is the index of refraction of the lens material, or the surface can be treated as indicated in Figure 6.8.

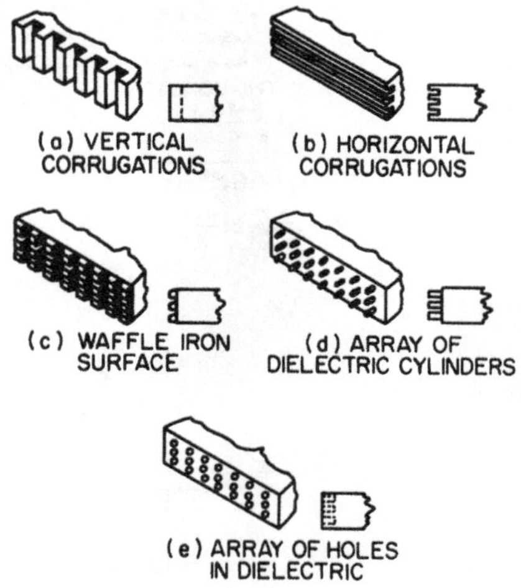

Figure 6.8 Simulated quarter-wavelength matching transformers for lens surfaces. (From Peeler [8], with permission of McGraw-Hill, Inc.)

REFERENCES

1. G.K. Huddleston and H.L. Bassett, "Radomes," Chapter 44 in *Antenna Engineering Handbook,* 2nd Ed. (R.C. Johnson and H. Jasik, eds.), New York: McGraw-Hill, 1984.
2. R.H. Cary, "Radomes," Chapter 14 in *Handbook of Antenna Design* (A.W. Rudge, K. Milne, A.D. Olver, and P. Knight, eds.), Vol. 2, London: Peter Peregrinus, 1983.
3. G.P. Tricoles, "Radome Electromagnetic Design," Chapter 31 in *Antenna Handbook* (Y.T. Lo and S.W. Lee, eds.), New York: Van Nostrand Reinhold, 1988.

4. J.D. Walton, ed., *Radome Engineering Handbook,* New York: Marcel Dekker, 1970.
5. H. Leaderman, "Electrical Design of Normal-Incidence Radomes," Chapter 10 in *Radar Scanners and Radomes* (W.M. Cady, M.B. Karelitz, and L.A. Turner, eds.), MIT Radiation Laboratory Series, Vol. 26, New York: McGraw-Hill, 1948.
6. H. Leaderman and L.A. Turner, "Theory of the Reflection and Transmission of Electromagnetic Waves by Dielectric Materials," Chapter 12 in *Radar Scanners and Radomes* (W.M. Cady, M.B. Karelitz, and L.A. Turner, eds.), MIT Radiation Laboratory Series, Vol. 26, New York: McGraw-Hill, 1948.
7. A.F. Kay, "Radomes and Absorbers," Chapter 32 in *Antenna Engineering Handbook* (H. Jasik, ed.), New York: McGraw-Hill, 1961.
8. G.D.M. Peeler, "Lens Antennas," Chapter 16 in *Antenna Engineering Handbook,* 2nd Ed. (R.C. Johnson and H. Jasik, eds.), New York: McGraw-Hill, 1984.
9. J.R. Risser, "Dielectric and Metal-plate Lenses," Chapter 11 in *Microwave Antenna Theory and Design* (S. Silver, ed.), MIT Radiation Laboratory Series, Vol. 12, New York: McGraw-Hill, 1949.
10. R.C. Johnson, "Optical Scanners," Chapter 3 in *Microwave Scanning Antennas* (R.C. Hansen, ed.), Vol. I, New York: Academic Press, 1964.
11. J.J. Lee, "Lens Antennas," Chapter 16 in *Antenna Handbook* (Y.T. Lo and S.W. Lee, eds.), New York: Van Nostrand Reinhold, 1988.
12. T. Milligan, "Lens Antennas," Chapter 9 in *Microwave Antenna Design,* New York: McGraw-Hill, 1985.

LIST OF SYMBOLS

A	variable defined in eq. (6.5)
D	thickness of dielectric core in A-sandwich
d	thickness of dielectric sheet or thickness of dielectric skins
k	variable defined in eq. (6.7)
$\|L\|^2$	power-loss coefficient
n	an integer such that D is positive [see eq. (6.11)]
$\|R\|^2$	power-reflection coefficient
$\|r_{ab}\|$	variable defined in eq. (6.9)
$\|T\|^2$	power-transmission coefficient
$\tan\delta$	loss tangent
ϵ	permittivity
ϵ_a, ϵ_b	permittivity of medium a and of medium b
ϵ_0	permittivity of free space
θ	angle of incidence (measured from normal to interface plane)
θ_B	angle of incidence for the Brewster phenomenon
κ	dielectric constant ($= \epsilon/\epsilon_0$).
κ_c	dielectric constant of core in A-sandwich radome
κ_{ce}	equivalent dielectric constant of core in A-sandwich radome
κ_s	dielectric constant of skins in A-sandwich radome
κ_{se}	equivalent dielectric constant of skins in A-sandwich radome
λ	wavelength in medium

λ_0	wavelength in free space
ϕ	variable defined in eq. (6.6) for solid-wall radomes and in eq. (6.14) for A-sandwich radomes
χ	variable defined in eq. (6.8)

Chapter 7
WAVEGUIDE COMPONENTS

In the design and fabrication of microwave antennas, we often must do some waveguide "plumbing." Normally, we would use off-the-shelf commercial components, but sometimes special requirements exist that dictate otherwise. In this chapter, we will discuss design techniques for some common waveguide components.

7.1 CORNERS

Corners for rectangular waveguides [1] are used for changing the direction of a waveguide run. The two main types are double-mitered corners and single-mitered cutoff corners as illustrated in Figure 7.1. The double-mitered corner may be easier to fabricate, and it can handle essentially the same peak power as straight waveguide. The single-mitered cutoff corner is smaller, but it can handle less power than straight waveguide—particularly in an E-plane corner.

7.1.1 Double-Mitered Corners

Note that the double-mitered corner of Figure 7.1(a) consists of two identical corners separated by a distance L measured along the central axis of the waveguide. We want the physical distance L to appear electrically to be one-quarter (or, on special occasions, three-quarters, five-quarters, *et cetera*) of a wavelength. Then, the two reflections will cancel and we will have a low VSWR corner.

For double-mitered E-plane corners, experiments have shown that L should be made equal to one-quarter of a guide wavelength in straight waveguide, or

$$L = \frac{\lambda_0/4}{\sqrt{1 - \left(\frac{\lambda_0}{2a}\right)^2}} \tag{7.1}$$

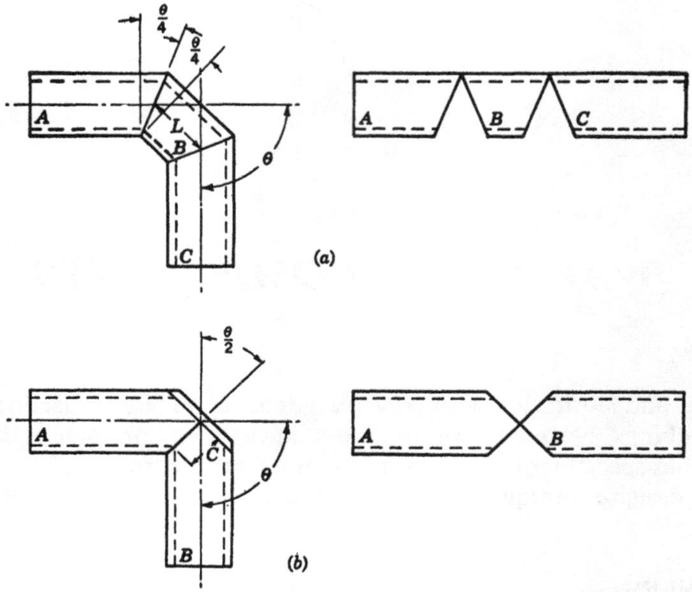

Figure 7.1 Waveguide corners: (a) double-mitered type; (b) single-mitered cutoff type. (From Ragan and Walker [1].)

For double-mitered H-plane corners, experiments have shown that L should be slightly larger than one-quarter of a guide wavelength, as indicated in Figure 7.2. These data were taken from measurements with 90-degree H-plane corners. For small values of the turn angle θ, L should be equal to one-quarter of a guide wavelength. For $0 < \theta < 90$ degrees, the designer will have to interpolate between one-quarter of a guide wavelength for $\theta = 0$ and the length indicated in Figure 7.2 for $\theta = 90$ degrees.

The typical bandwidth for a 90-degree double-mitered H-plane corner is illustrated in Figure 7.3. Note that this is satisfactory for most radar applications.

7.1.2 Single-Mitered Cutoff Corners

A single-mitered cutoff corner is illustrated in Figure 7.1(b). The designer must determine the appropriate value for the dimension C—the distance from the inside corner to the cutoff plate. Data from Ragan and Walker [1] can be arranged to present C/b for E-plane corners and C/a for H-plane corners as functions of the turn angle θ in Figure 7.4. Note that, for a particular waveguide, C is given as a

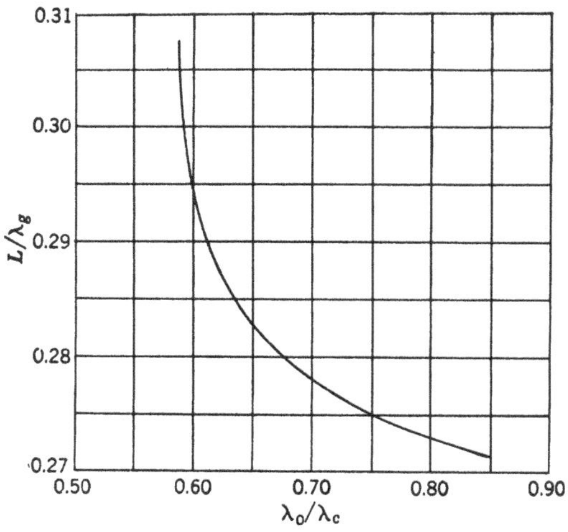

Figure 7.2 Design curve for 90-degree double-mitered H-plane corners. (From Ragan and Walker [1].)

function of θ but not of frequency. The impedance match is broad-banded and usually satisfactory across the waveguide bandwidth; however, if it is not, the designer will have to make some adjustments to achieve the required match.

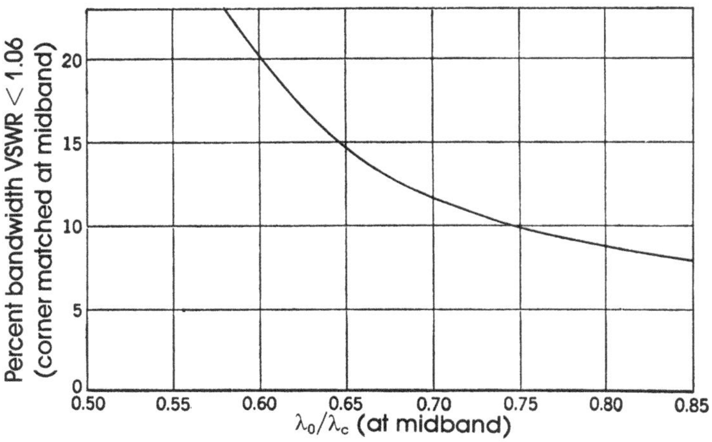

Figure 7.3 Bandwidth curve for 90-degree double-mitered H-plane corners. (From Ragan and Walker [1].)

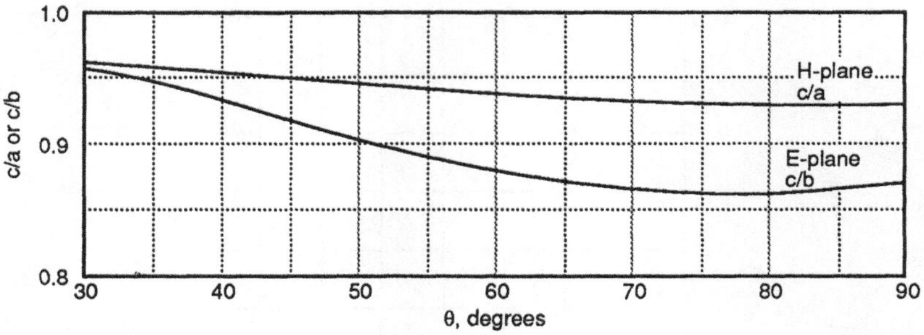

Figure 7.4 Design curves for single-mitered cutoff corners.

7.2 BENDS

Circular bends (constant radius) are smooth corners as illustrated in Figure 7.5. The bend acts as a transformer, setting up reflections of equal magnitude at each end. The transformation is from a higher to lower impedance at the initial end and from a lower to higher impedance at the other end, or *vice versa*. Therefore, the electrical length of the bend should be any number of half wavelengths long measured along the waveguide axis to cancel the reflections.

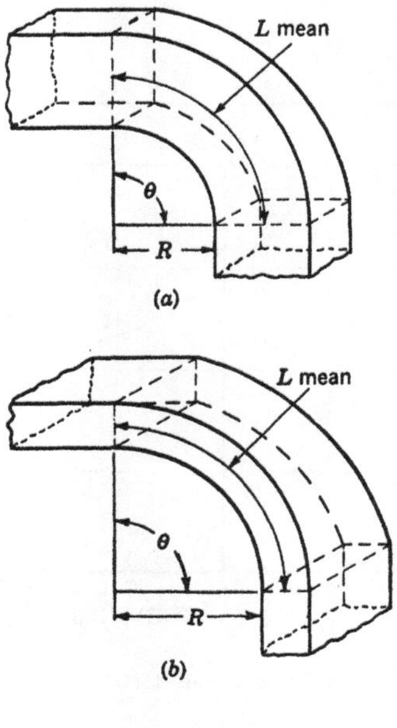

Figure 7.5 Types of waveguide bends: (a) H-plane; (b) E-plane.

Rice [2] derived equations for the propagation constant in bends, and Alison [3] presented data in a useful manner for microwave designers. Starting with E-plane bends, the absolute magnitude of the reflection coefficient is

$$|\Gamma| = \frac{1}{24}\left(\frac{b}{R}\right)^2 \left|\sin\left(\frac{2\pi L}{\lambda_g}\right) - A\cos\left(\frac{2\pi L}{\lambda_g}\right)\right| \tag{7.2}$$

for $R/b > 2$ and where

$$A = \frac{96}{\pi^4}\left(\frac{2b}{\lambda_g}\right)\sum_{1,3}^{\infty} m^{-5}\left[1 - \left(\frac{2b}{m\lambda_g}\right)^2\right]^{-1/2} \tag{7.3}$$

The condition for a match is

$$\frac{2\pi L}{\lambda_g} = \arctan A \tag{7.4}$$

or

$$R = \frac{\lambda_g}{2\pi\theta}\arctan A \tag{7.5}$$

The first root of eq. (7.4) is plotted in Figure 7.6; the length L should be increased in steps of $\lambda_g/2$ until a suitable value of R is obtained. To be precise, λ_g

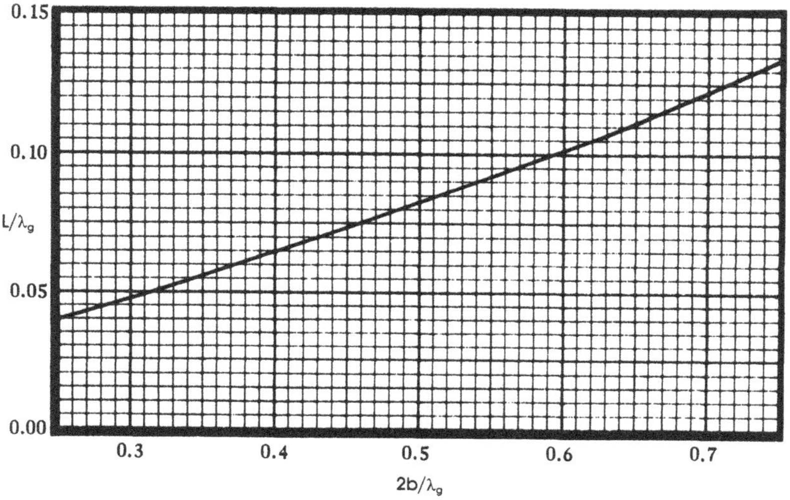

Figure 7.6 Design curve for E-plane bends. (From Alison [3].)

should be the guide wavelength in the bend, but a much simpler solution (and one that yields satisfactory results) is to let λ_g be the guide wavelength in straight waveguide.

For example, suppose we want to design an E-plane 90-degree bend in WR-90 waveguide (a = 0.900 and b = 0.400 in.) for a design frequency of 10 GHz. We have λ_0 = 1.180 in., λ_g = 1.563 in., and $2b/\lambda_g$ = 0.512. From Figure 7.6, $L = 0.085\lambda_g$ = 0.133 in.; thus, R = 0.085 in. This is physically impossible, so add a half guide-wavelength to L. Then, L = 0.915 in. and R = 0.583 in.

If desired, the value of A can be obtained from the graph in Figure 7.7, which is useful if we manually calculate the reflection coefficient *versus* frequency using eq. (7.2).

For H-plane bends,

$$|\Gamma| = \frac{\lambda_g^2}{32\pi^2 R^2} \left| \sin\left(\frac{2\pi L}{\lambda_g}\right) - \frac{128}{\pi^2}\frac{a}{\lambda_g} B \cos\left(\frac{2\pi L}{\lambda_g}\right) \right| \tag{7.6}$$

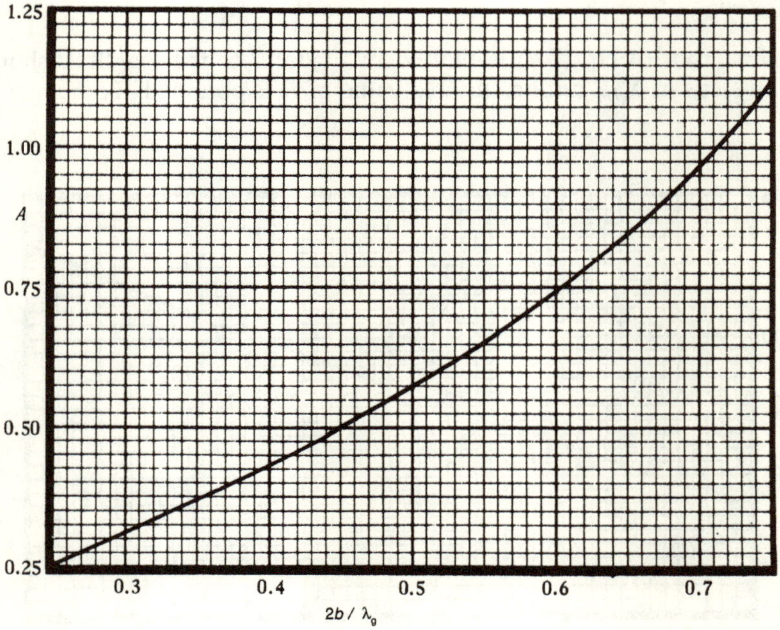

Figure 7.7 Curve for value of A. (From Alison [3].)

where

$$B = \sum_{2,4}^{\infty} m^2(m^2 - 1)^{-3} \left[(m^2 - 1) - \left(\frac{2a}{\lambda_g} \right)^2 \right]^{-1/2} \tag{7.7}$$

The condition for a match is

$$L = \frac{\lambda_g}{2\pi} \arctan \left(\frac{128}{\pi^2} \frac{a}{\lambda_g} B \right) \tag{7.8}$$

or

$$R = \frac{\lambda_g}{2\pi\theta} \arctan \left[\frac{128}{\pi^2} \frac{a}{\lambda_g} B \right] \tag{7.9}$$

The first root of eq. (7.8) is plotted in Figure 7.8; the length L should be increased in steps of $\lambda_g/2$ until a suitable value of R is obtained. If desired, the value of B can be obtained from the graph in Figure 7.9. The curve of Figure 7.8 allows us to design quickly a matched bend using the input parameters of waveguide size, frequency, and turn angle. The curve of Figure 7.9 allows us to determine quickly the value of B for manually calculating the reflection coefficient *versus* frequency using eq. (7.6).

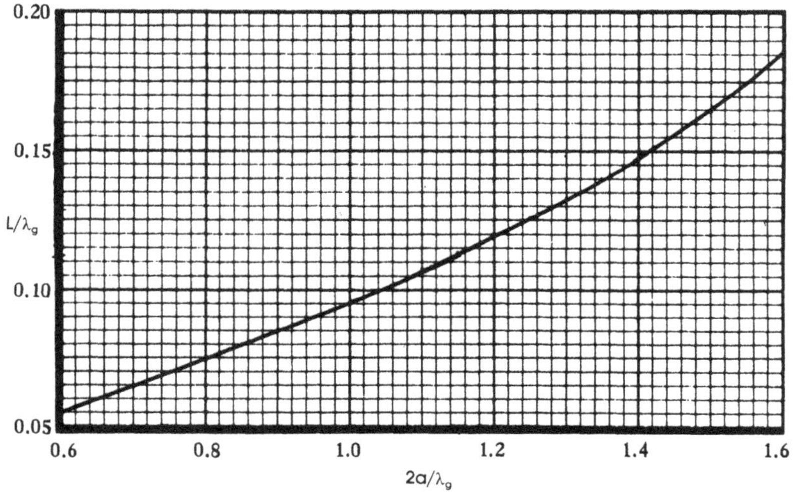

Figure 7.8 Design curve for H-plane bends. (From Alison [3].)

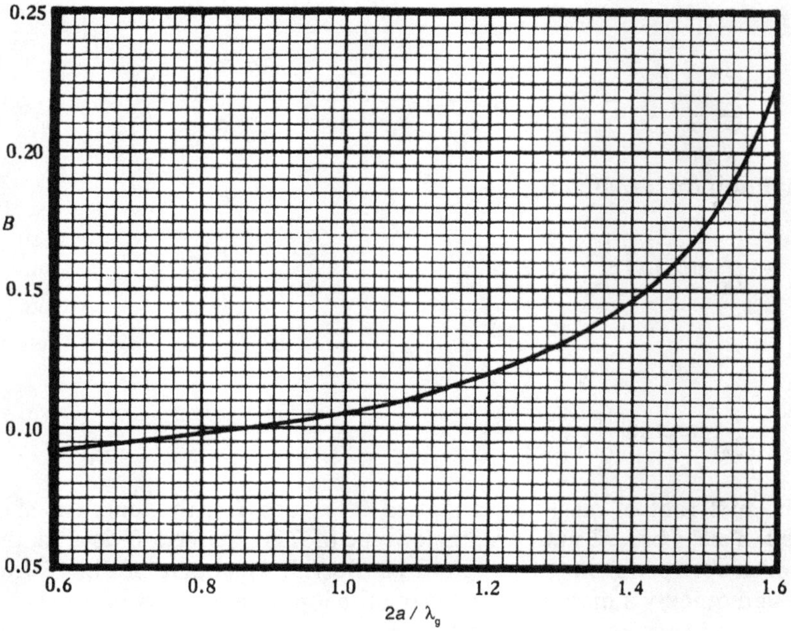

Figure 7.9 Curve for value of *B*. (From Alison [3].)

For example, suppose that we want to design an H-plane 90-degree bend in WR-90 waveguide (a = 0.900 and b = 0.400 in.) for a design frequency of 10 GHz. We have λ_0 = 1.180 in., λ_g = 1.563 in., and $2a/\lambda_g$ = 1.152. From Figure 7.8, $L = 0.113\lambda_g$ = 0.177 in.; thus, R = 0.113 in. Again, this is physically impossible, so add a half guide-wavelength to L. Then, L = 0.959 in. and R = 0.611 in.

7.3 TWISTS

Twists are useful components for changing the polarization of a waveguide (i.e., rotate the cross section of the waveguide). Figure 7.10 illustrates both a step twist and a smooth twist. Twists are difficult to fabricate so off-the-shelf commercial models are used when possible.

Step twists [4, 5] consist of a series of quarter-wavelength sections that are stepped in angle in a manner to reduce the resultant reflection from the component. Each field reflection coefficient is proportional to the square of the angle

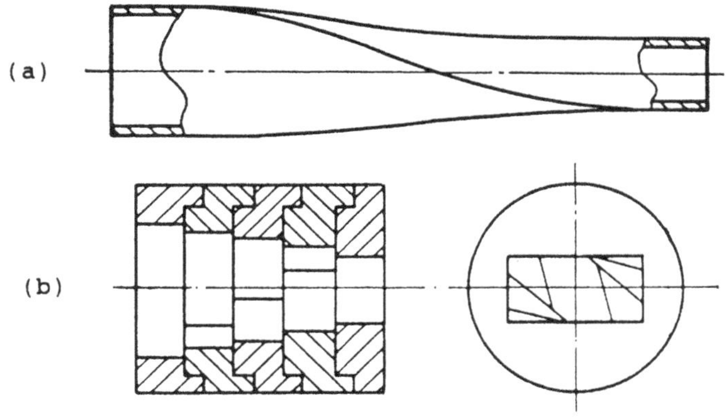

Figure 7.10 Twists in rectangular waveguides: (a) smooth twist; (b) step twist.

between sections, so the angles are arranged such that the reflection amplitudes follow a binomial series. The bandwidth increases with the number of sections; an example of a three-section twist had a VSWR of less than 1.05 over a 10% bandwidth.

Smooth twist waveguides were analyzed by Lewin [6] and a design procedure was presented by Alison [3]. The wavelength can be expressed as

$$\frac{\lambda_t}{\lambda_g} = \left[1 - \frac{ha^2}{l^2}\right]^{-1/2} \tag{7.10}$$

where

λ_t = wavelength in twist
λ_g = wavelength in straight waveguide
l = length required to rotate waveguide through 90 degrees
h = a function of other waveguide constants (see Figure 7.11)
a = width of the waveguide

The impedances are also related as

$$\frac{Z_t}{Z_0} = \left[1 - \frac{ha^2}{l^2}\right]^{-1/2} \tag{7.11}$$

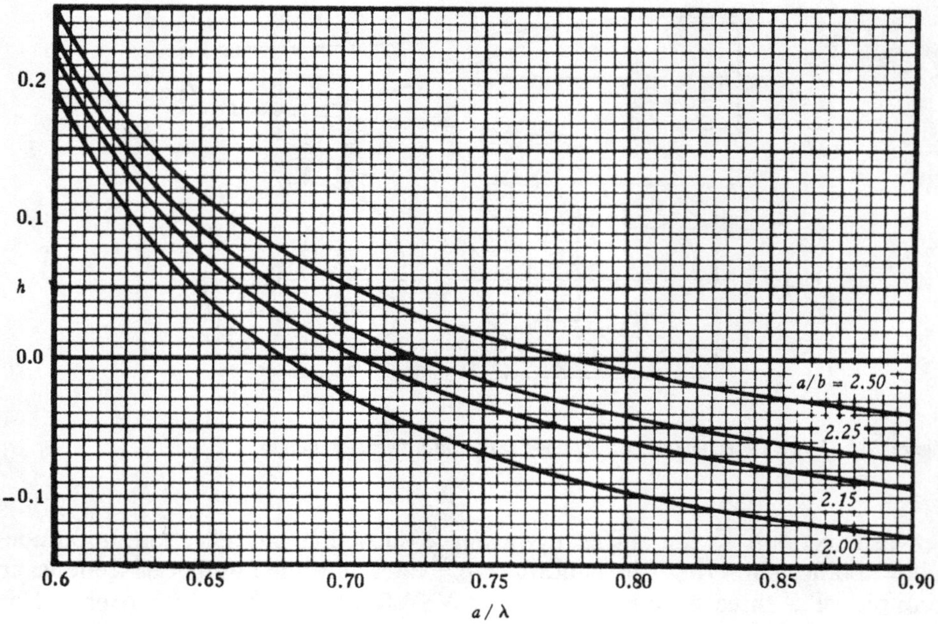

Figure 7.11 Value of *h versus* a/λ for various a/b ratios. (From Alison [3].)

The twist will have two main reflections: one at the beginning of the twist and one at the end of the twist. The absolute magnitudes of the two reflection coefficients are

$$|\Gamma_i| = |\Gamma_o| = \left|\frac{Z_t - Z_0}{Z_t + Z_0}\right| \tag{7.12}$$

The total reflection coefficient is

$$\Gamma = |\Gamma_i| - |\Gamma_o| \left[\cos\left(4\pi \frac{x}{\lambda_t}\right) + j \sin\left(4\pi \frac{x}{\lambda_t}\right)\right] \tag{7.13}$$

where x is the distance from the input end of the twisting waveguide. Notice that the reflection coefficient is zero when the length is even multiples of a quarter wavelength, and it is maximum at odd multiples of a quarter wavelength.

The VSWR is

$$\text{VSWR} = \frac{1 + |\Gamma|}{1 - |\Gamma|} \tag{7.14}$$

7.4 TAPERS

Occasionally, we need to transform from one size of rectangular waveguide to another size. Usually, a stepped transformer or a taper is used. The taper profile can be linear, sinusoidal, exponential, *et cetera.* The linear taper [7, 8] will be discussed here because it is simple to design and fabricate and it performs almost as well as the more complex designs.

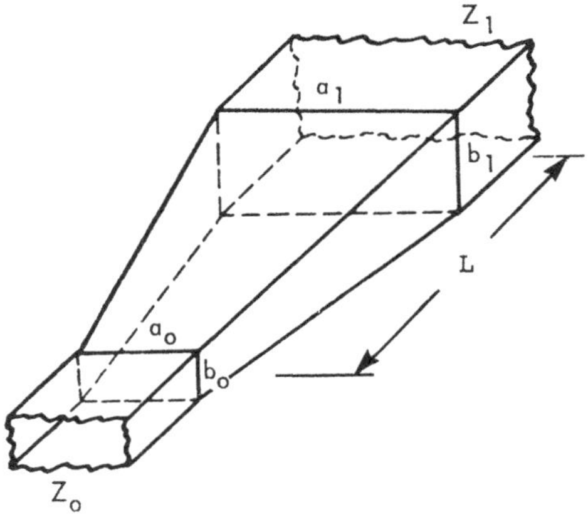

Figure 7.12 Linear taper of axial length L connecting rectangular waveguides of impedances Z_0 and Z_1.

Figure 7.12 illustrates the parameters of a linear taper. The absolute magnitude of the reflection coefficient is

$$|\Gamma| = \frac{1}{L/\lambda_0}\left[\frac{K_0^2 + K_1^2}{64\pi^2} - \frac{K_0 K_1}{32\pi^2}\cos(4\pi l)\right]^{1/2} \tag{7.15}$$

where

$$K_0 = \frac{\dfrac{b_1 - b_0}{b_0} - \dfrac{a_1 - a_0}{a_0}\left[\dfrac{\kappa}{\kappa - (\lambda_0/2a_0)^2}\right]}{[\kappa - (\lambda_0/2a_0)^2]^{1/2}} \tag{7.16}$$

$$K_1 = \frac{\dfrac{b_1 - b_0}{b_1} - \dfrac{a_1 - a_0}{a_1}\left[\dfrac{\kappa}{\kappa - (\lambda_0/2a_1)^2}\right]}{[\kappa - (\lambda_0/2a_1)^2]^{1/2}} \tag{7.17}$$

$$l = \frac{L}{2(a_1 - a_0)}\left(\frac{2a_1}{\lambda_{g1}} - \frac{2a_0}{\lambda_{g0}} + \arctan\frac{2a_0}{\lambda_{g0}} - \arctan\frac{2a_1}{\lambda_{g1}}\right) \tag{7.18}$$

$$\lambda_{g0} = \lambda_0/[\kappa - (\lambda_0/2a_0)^2]^{1/2} \tag{7.19}$$

$$\lambda_{g1} = \lambda_0/[\kappa - (\lambda_0/2a_1)^2]^{1/2} \tag{7.20}$$

In the above equations, κ is the dielectric constant of the medium inside the waveguide (for air, set $\kappa = 1$). At first glance, the equations seem formidable, but they are simple and easy to evaluate on a computer or programmable hand-held calculator. The VSWR can be calculated from eq. (7.14).

REFERENCES

1. G.L. Ragan and R.M. Walker, "Rigid Transmission Lines," Chapter 4 in *Microwave Transmission Circuits* (G.L. Ragan, ed.), MIT Radiation Laboratory Series, Vol. 9, New York: McGraw-Hill, 1948.
2. S.O. Rice, "Reflection from Circular Bends in Rectangular Waveguide," *Bell Syst. Tech. J.*, 1948, pp. 305–349.
3. W.B.W. Alison, *A Handbook for the Mechanical Tolerancing of Waveguide Components,* Norwood, MA: Artech House, 1987, Chap. 3.
4. A.F. Harvey, *Microwave Engineering,* New York: Academic Press, 1963, Sec. 2.6.1.
5. H.A. Wheeler and H. Schwiebert, "Step-Twist Waveguide Components," *Trans. IRE,* Vol. MTT-3, 1955, p. 45.
6. L. Lewin, "Propagation in Curved and Twisted Waveguides of Rectangular Cross-Section," *Proc. IEE,* Part B, Vol. 102, 1955, pp. 75–80.
7. R.C. Johnson, "Design of Linear Double Tapers in Rectangular Waveguides," *IEEE Trans. Microwave Theory Tech.,* Vol. MTT-7, July 1959, pp. 374–378. (Corrections: Vol. MTT-8, July 1960, p. 458.)
8. R.C. Johnson and D.J. Bryant, "Linear Tapers in Rectangular Waveguides," *IEEE Trans. Microwave Theory Tech.,* Vol. MTT-9, May 1961, p. 261.

LIST OF SYMBOLS

A variable defined in eq. (7.3)

a inside width of rectangular waveguide

a_0	inside width of rectangular waveguide at input to a linear taper
a_1	inside width of rectangular waveguide at output of a linear taper
B	variable defined in eq. (7.7)
b	inside height of rectangular waveguide
b_0	inside height of rectangular waveguide at input to a linear taper
b_1	inside height of rectangular waveguide at output of a linear taper
c	distance from the inside corner to the cutoff plate in a single-mitered cutoff corner
h	a function of other waveguide constants (see Figure 7.11)
K_0	dummy variable defined in eq. (7.16)
K_1	dummy variable defined in eq. (7.17)
L	separation (measured along waveguide axis) between miters of a double-mitered corner, length of a bend measured along the mean radius, or axial length of a linear waveguide taper
l	length of smooth twist required to rotate waveguide through 90 degrees or electrical length of a linear waveguide taper
m	dummy integer variable in a summation operation
R	mean radius of a bend
VSWR	voltage standing wave ratio
x	distance from the input end of a twisting waveguide
Z_0	characteristic impedance of rectangular waveguide
Z_t	characteristic impedance of twisted rectangular waveguide
Γ	reflection coefficient
Γ_i	reflection coefficient at the input to a twisted waveguide
Γ_o	reflection coefficient at the output from a twisted waveguide
θ	turn angle through which a corner changes direction of the waveguide
κ	dielectric constant
λ_0	wavelength in free space
λ_c	cutoff wavelength in waveguide
λ_g	wavelength in waveguide
λ_{g0}	guide wavelength at input to linear waveguide taper
λ_{g1}	guide wavelength at output of linear waveguide taper
λ_t	wavelength in a smooth twist waveguide

Chapter 8
IMPEDANCE AND MATCHING

This chapter will discuss basic notations, reflections from mismatches, impedance measurements, and matching with known susceptances of simple circuit elements. The subject is discussed more thoroughly in [1–6].

8.1 BASIC NOTATION

Impedance and related characteristics are usually represented by complex numbers in mathematical expressions. Let

- Z = impedance
- R = resistance
- X = reactance

Then,

$$Z = R + jX \tag{8.1}$$

Resistance is the real component and reactance is the imaginary component of impedance. If X is positive, it is inductive; if X is negative, it is capacitive.

The reciprocal of impedance is admittance. Let

- Y = admittance
- G = conductance
- B = susceptance

Then,

$$Y = G + jB \tag{8.2}$$

and

$$Y = 1/Z \qquad \text{and} \qquad Z = 1/Y \tag{8.3}$$

Conductance is the real component and susceptance is the imaginary component of admittance. If B is positive, it is capacitive; if B is negative, it is inductive. The term "immittance" is a general term used sometimes to represent either impedance or admittance.

The above terms are shown in the real-imaginary plane in Figure 8.1. The rectangular components can be converted to an exponential form using

$$e^{j\theta} = \cos\theta + j\sin\theta \tag{8.4}$$

For example,

$$Z\, e^{j\theta} = Z\cos\theta + jZ\sin\theta = R + jX \tag{8.5}$$

where

$$\theta = \arctan(X/R) \tag{8.6}$$

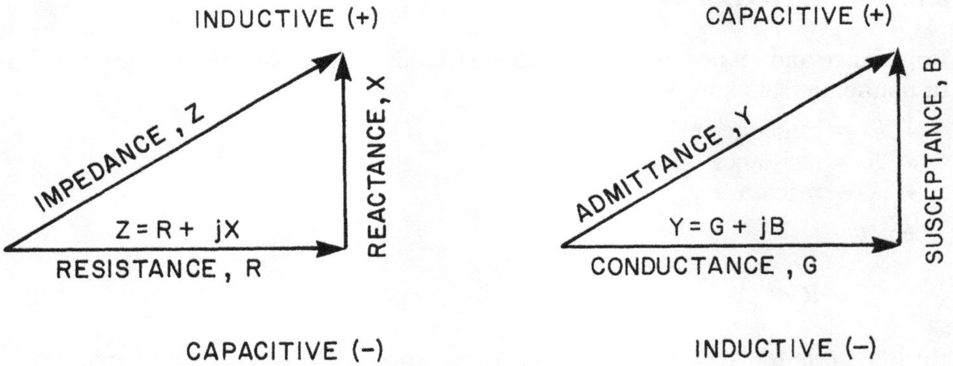

Figure 8.1 Basic notations for complex circuit parameters.

8.2 REFLECTIONS FROM MISMATCHES

Suppose that a transmission line is terminated with an impedance that is not equal to the characteristic impedance of the transmission line. Let

- Γ = voltage reflection coefficient
- VSWR = voltage standing wave ratio
- Z_0 = characteristic impedance of transmission line
- Z_L = load impedance

Then,

$$\Gamma = \frac{Z_L - Z_0}{Z_L + Z_0} \tag{8.7}$$

$$\text{VSWR} = \frac{1 + |\Gamma|}{1 - |\Gamma|} = \frac{|Z_L|}{|Z_0|} \tag{8.8}$$

$$|\Gamma| = \frac{\text{VSWR} - 1}{\text{VSWR} + 1} \tag{8.9}$$

Suppose that a lossless transmission line has a discontinuity; portions of an incident wave will be reflected by or transmitted past the discontinuity. Let

- V_i, V_r, V_t = voltage (magnitudes) of incident, reflected, and transmitted waves, respectively
- L_r, L_t = return loss and transmission loss, respectively, dB

Then,

$$L_r = -10 \log \frac{V_r^2}{V_i^2} = -20 \log|\Gamma| \tag{8.10}$$

$$L_t = -10 \log \frac{V_t^2}{V_i^2} = -10 \log \frac{V_i^2 - V_r^2}{V_i^2}$$

$$= -10 \log(1 - |\Gamma|^2) \tag{8.11}$$

A standing wave will reduce the peak power-handling capability of a transmission line. Let

- P_{mo} = maximum power transmission on a matched line
- P_{ms} = maximum power transmission on a line having a standing wave

Then (p. 32 [3]),

$$P_{ms} = \frac{P_{mo}}{\text{VSWR}} \tag{8.12}$$

8.3 IMPEDANCE MEASUREMENTS AND THE SMITH CHART

If a network analyzer and experienced operator are available, they normally will be used to make impedance measurements because such measurements can be made quickly and accurately. In some cases, however, a slotted line must be used. The discussion in this chapter will be based on slotted-line measurements because they are simple to make and easy to understand.

The impedance can be measured with a slotted line by observing the standing wave patterns such as those in Figure 8.2. First, put the load on the line, measure the VSWR, and note the positions of the minima. Second, insert a short in the line. Observe how the positions of the minima (or nulls) have shifted, and measure the guide wavelength (which is equal to twice the distance between two adjacent minima). The shift in the positions of the minima yields phase information for the load impedance.

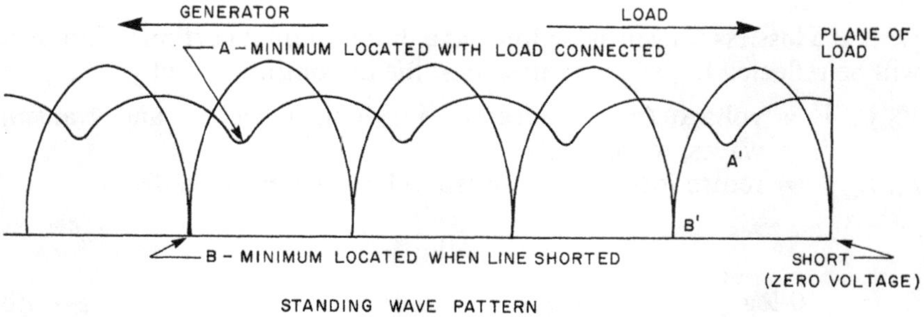

Figure 8.2 Standing wave patterns on a slotted line.

The following statements are helpful when interpreting slotted-line impedance measurements:

1. The shifts in minima positions when the load is shorted are never more than $\lambda_g/4$.
2. If shorting the load causes the minima positions to shift toward the load, a capacitive component exists in the load.
3. If shorting the load causes the minima positions to shift toward the generator, an inductive component exists in the load.
4. If shorting the load causes no shift in the minima positions, the load is completely resistive with a magnitude of Z_0/VSWR.
5. If shorting the load causes the minima positions to shift exactly $\lambda_g/4$, the load is completely resistive with a magnitude of (Z_0)(VSWR).

The effects of these statements are illustrated in Figure 8.3. The impedance we measure will be that at the position of the short.

The Smith chart [7–9] is the most popular chart for recording impedance and admittance data (which are normalized to the characteristic impedance or admittance of the transmission line). To simplify this discussion, we will describe it as an impedance chart.

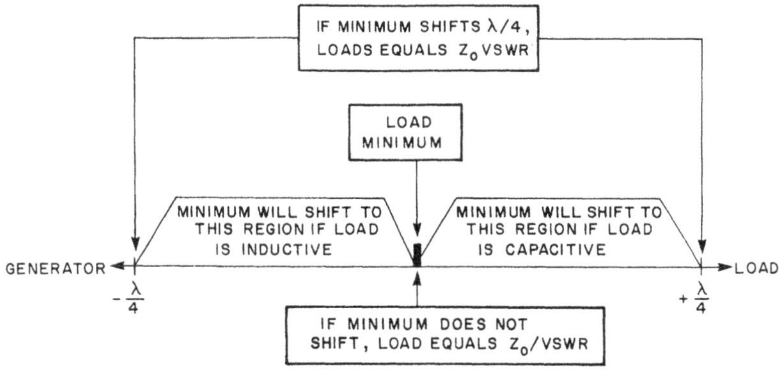

Figure 8.3 Interpreting standing wave patterns. (In this figure, λ is guide wavelength.)

The Smith chart is a circular chart as illustrated in Figure 8.4; at the center, $R = 1$ and $X = 0$. The radial distance from the center represents $|\Gamma|$ on a linear scale from 0 at the center to 1.0 at the edge (the unity circle). Circles that are concentric with the unity circle are circles of constant magnitude of the reflection coefficient (and also constant VSWR). Some of the important characteristics are summarized as follows:

1. The constant R and constant X loci form two families of orthogonal circles in the chart.
2. All of the constant R and constant X circles pass through the point on the right side where the real axis intersects the unity circle.
3. The upper half of the chart represents $+jX$ and the lower half represents $-jX$.
4. For admittance, the constant R circles become constant G circles, and the constant X circles become constant B circles.
5. The distance around the Smith chart once is one-half of a guide wavelength.
6. The horizontal radius to the right of the chart center corresponds to maximum voltage, minimum current, maximum impedance, and VSWR.
7. The horizontal radius to the left of the chart center corresponds to minimum voltage, maximum current, minimum impedance, and 1/VSWR.
8. The normalized admittance is diametrically opposite the normalized impedance (i.e., 180-degree phase difference and an equal distance from the chart center).
9. Any physically realizable passive impedance plotted on a Smith chart displays a circular or spiral motion having a clockwise sense of rotation with frequency [5].

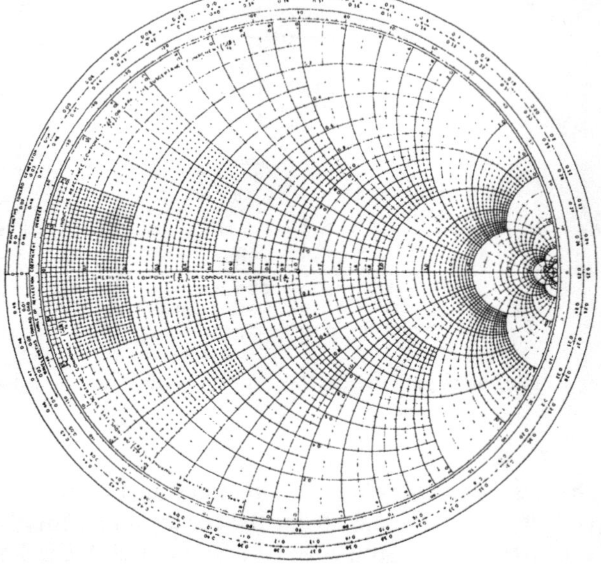

Figure 8.4 The Smith chart.

We now proceed to a sample problem. Suppose that we are working with WR-90 waveguide (a = 0.900 and b = 0.400 in.) at a frequency of 10 GHz. The load has a VSWR = 1.55. When the load is shorted, the minima shift 0.15 in. toward the generator; the guide wavelength is 1.563 in. The shift in guide wavelengths is 0.15/1.563 = 0.096. The impedance at the position of the short can be plotted as point A on a Smith chart as indicated in Figure 8.5. This sample problem will be continued at the end of the next section.

8.4 CIRCUIT ELEMENTS FOR MATCHING

Several simple circuit elements can be used for matching; this chapter will present susceptance data for some thin irises and posts [10]. Figure 8.6 presents susceptance data for a single inductive post (parallel to the E-field). Figure 8.7 presents similar data for a symmetric inductive post doublet. Figures 8.8 and 8.9 present susceptance data for thin single and double inductive irises. Figure 8.10 presents susceptance data for various capacitive irises. Design data for other circuit elements are available in the references and in other sources.

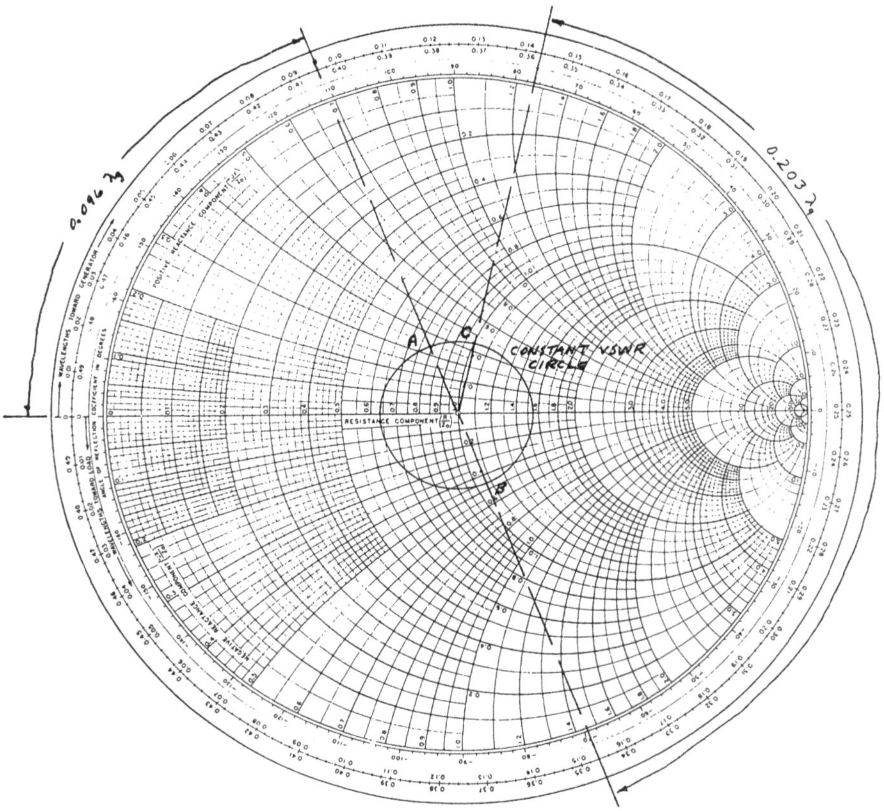

Figure 8.5 Sample problem using a Smith chart.

Usually, inductive circuit elements are preferred in radar applications because capacitive elements often reduce the peak power capabilities of the waveguide.

Now we continue the sample problem of the previous section. We had just plotted the impedance as point A of Figure 8.5. We decide to match the load with a single inductive post. Such a post is a shunt element, and its susceptance is given in Figure 8.6; thus, we need to convert from impedance to admittance in our Smith chart. Continue the line from point A through the center of the chart; point B, where this line intersects the constant VSWR circle, is the admittance. Now rotate around the chart 0.203 guide wavelengths (0.317 in.) toward the load to point C where the admittance is $1 + j0.45$. We want to insert an inductive post with a susceptance of -0.45 at this position in the waveguide.

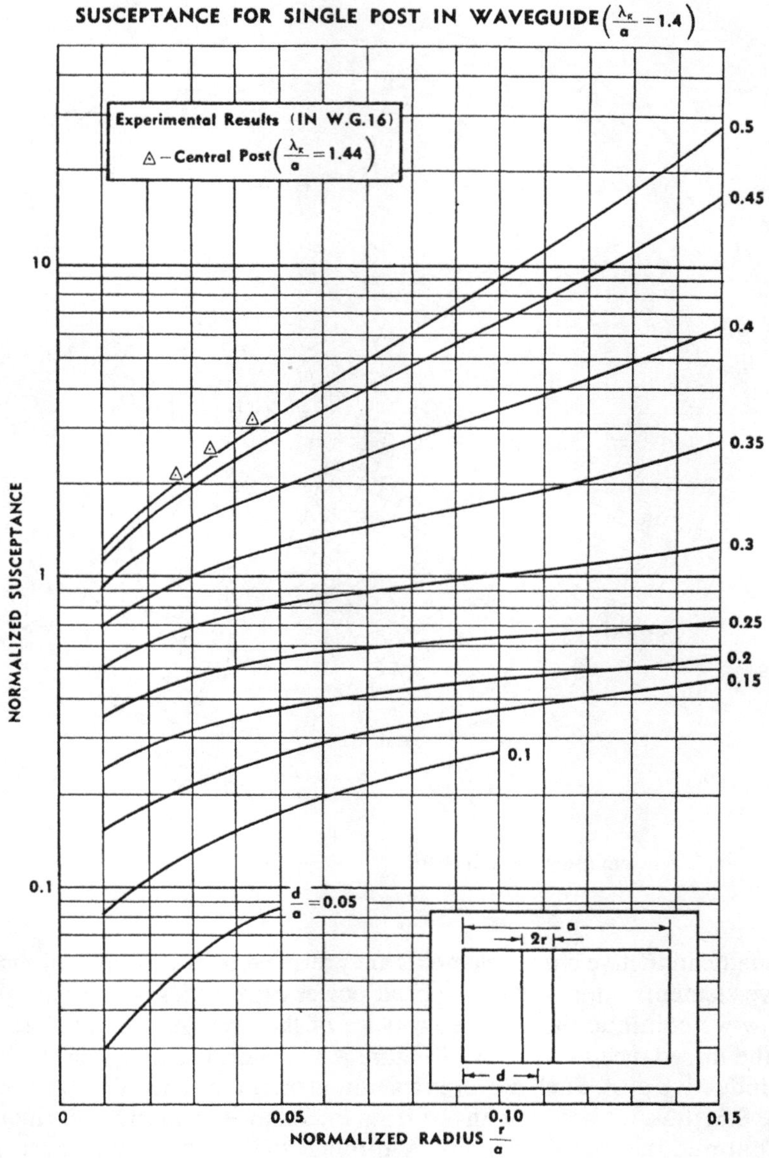

Figure 8.6(a) Susceptance for single inductive post in rectangular waveguide when $\lambda_g/a = 1.4$. (From Saad [10].)

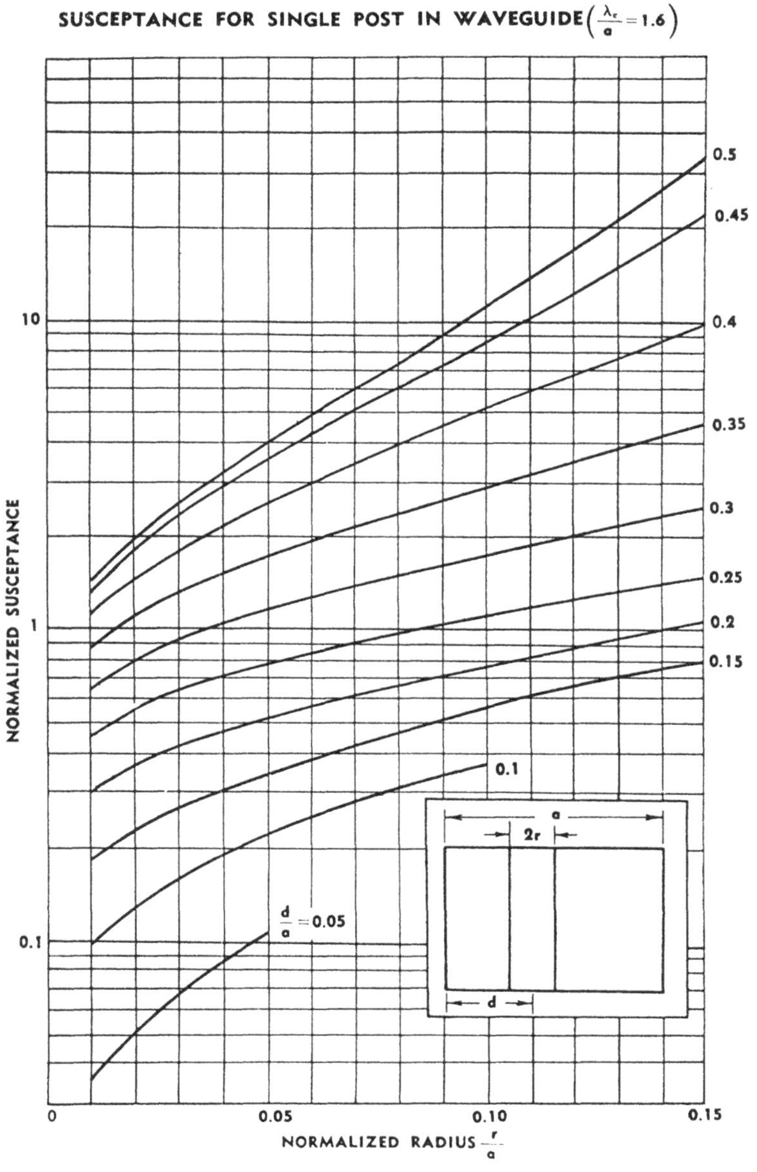

Figure 8.6(b) Susceptance for single inductive post in rectangular waveguide when λ_g/a = 1.6. (From Saad [10].)

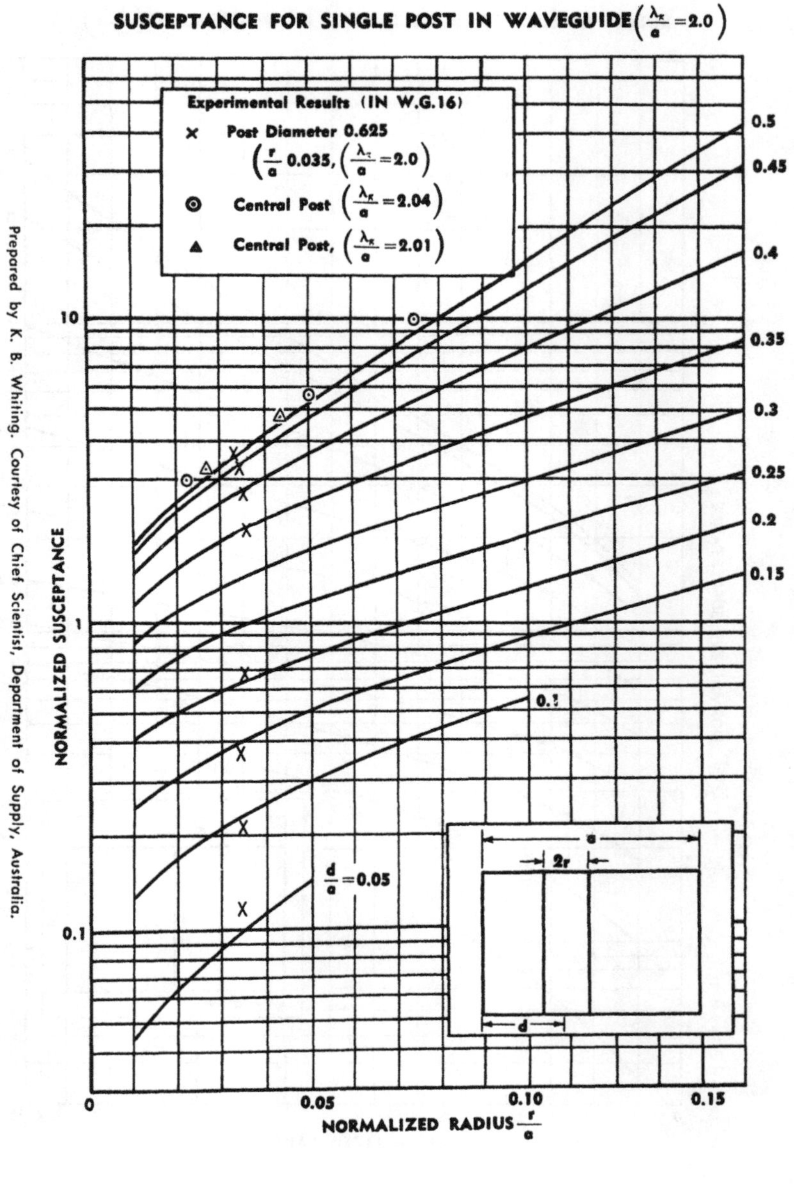

Figure 8.6(c) Susceptance for single inductive post in rectangular waveguide when $\lambda_g/a = 2.0$. (From Saad [10].)

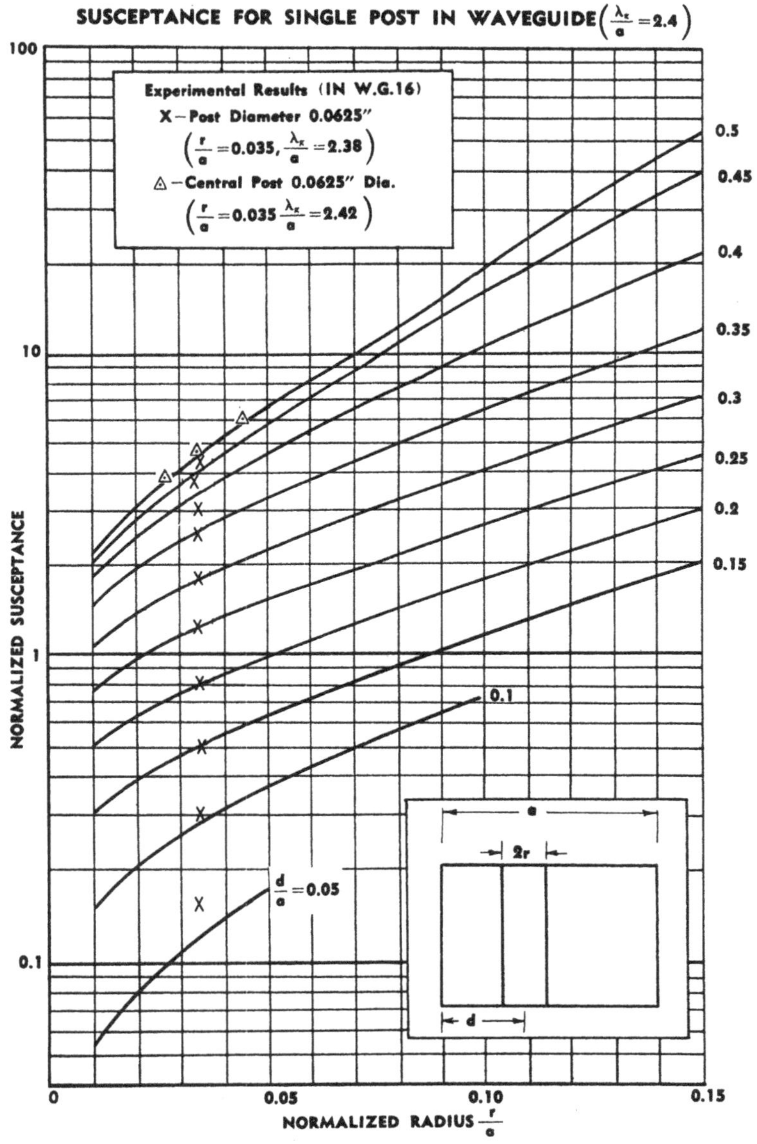

Figure 8.6(d) Susceptance for single inductive post in rectangular waveguide when $\lambda_g/a = 2.4$. (From Saad [10].)

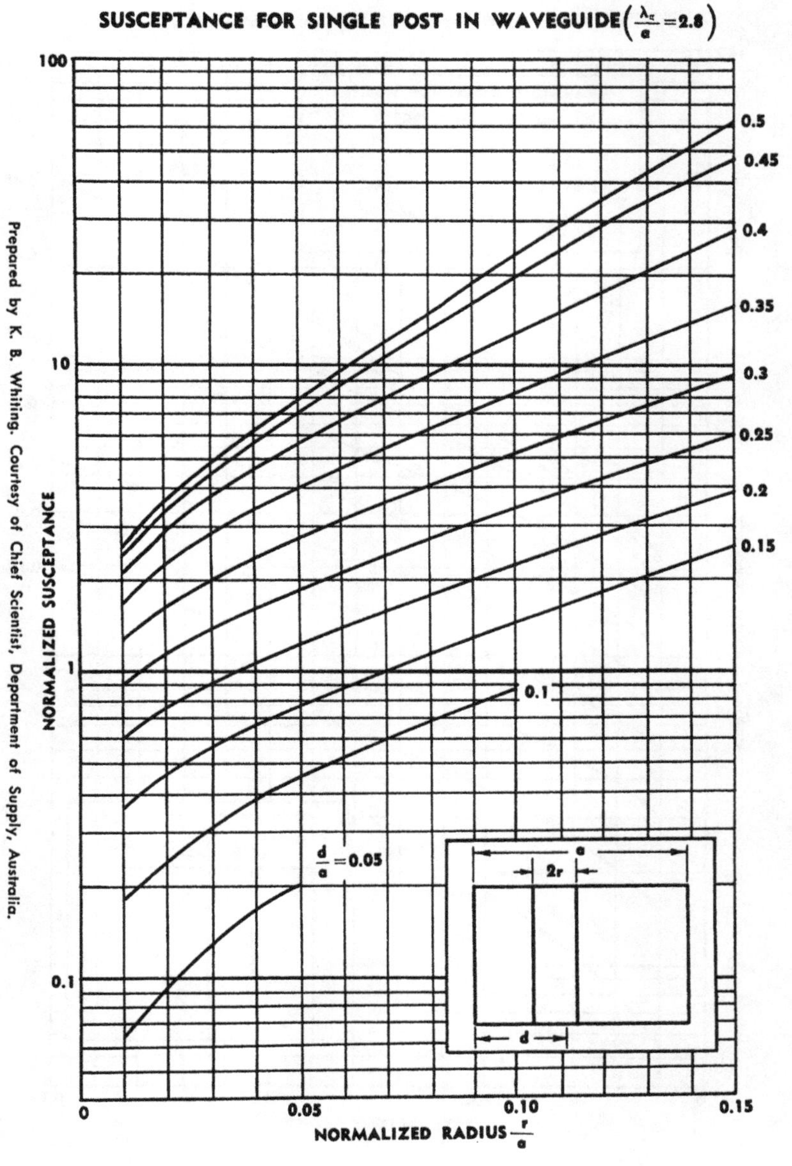

Figure 8.6(e) Susceptance for single inductive post in rectangular waveguide when $\lambda_g/a = 2.8$. (From Saad [10].)

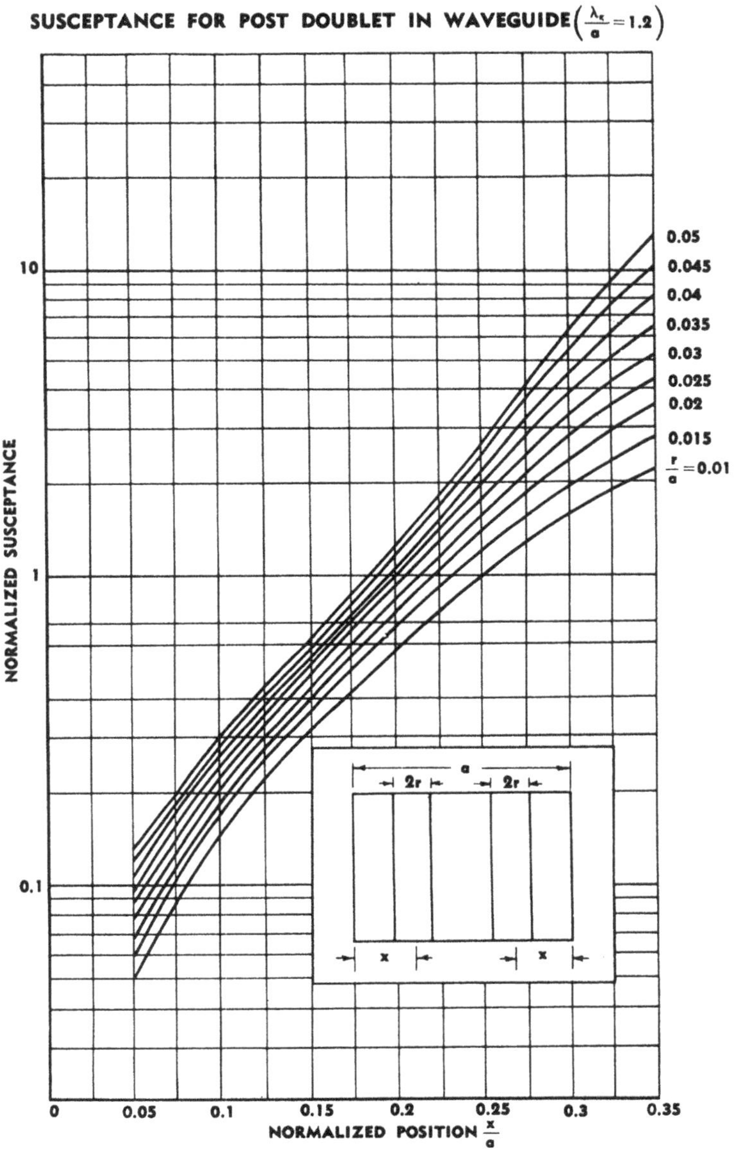

Figure 8.7(a) Susceptance for inductive post doublet in rectangular waveguide when $\lambda_g/a = 1.2$. (From Saad [10].)

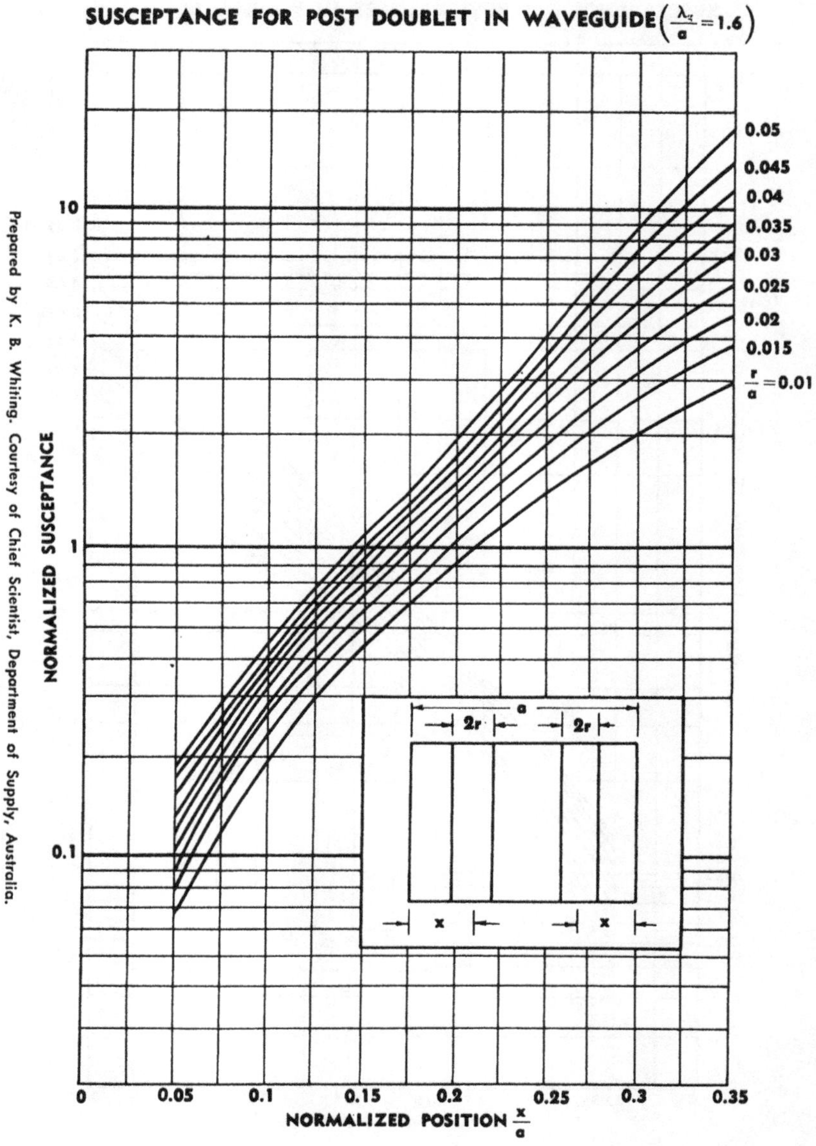

Figure 8.7(b) Susceptance for inductive post doublet in rectangular waveguide when $\lambda_g/a = 1.6$. (From Saad [10].)

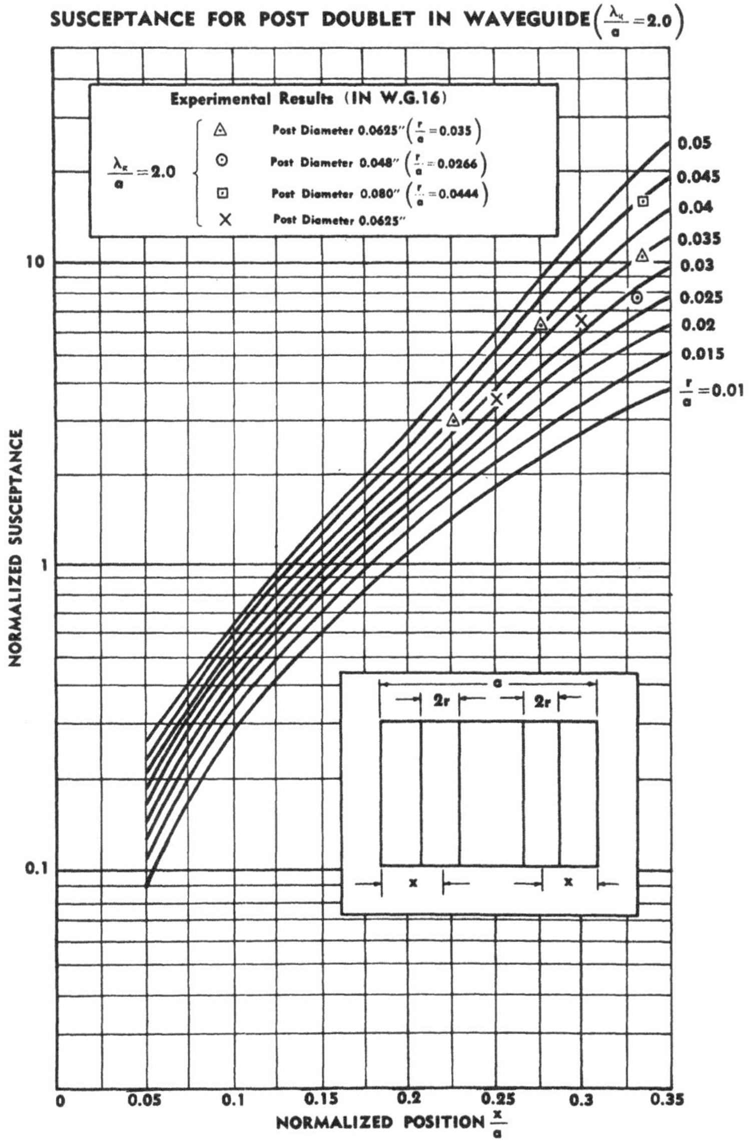

Figure 8.7(c) Susceptance for inductive post doublet in rectangular waveguide when $\lambda_g/a = 2.0$. (From Saad [10].)

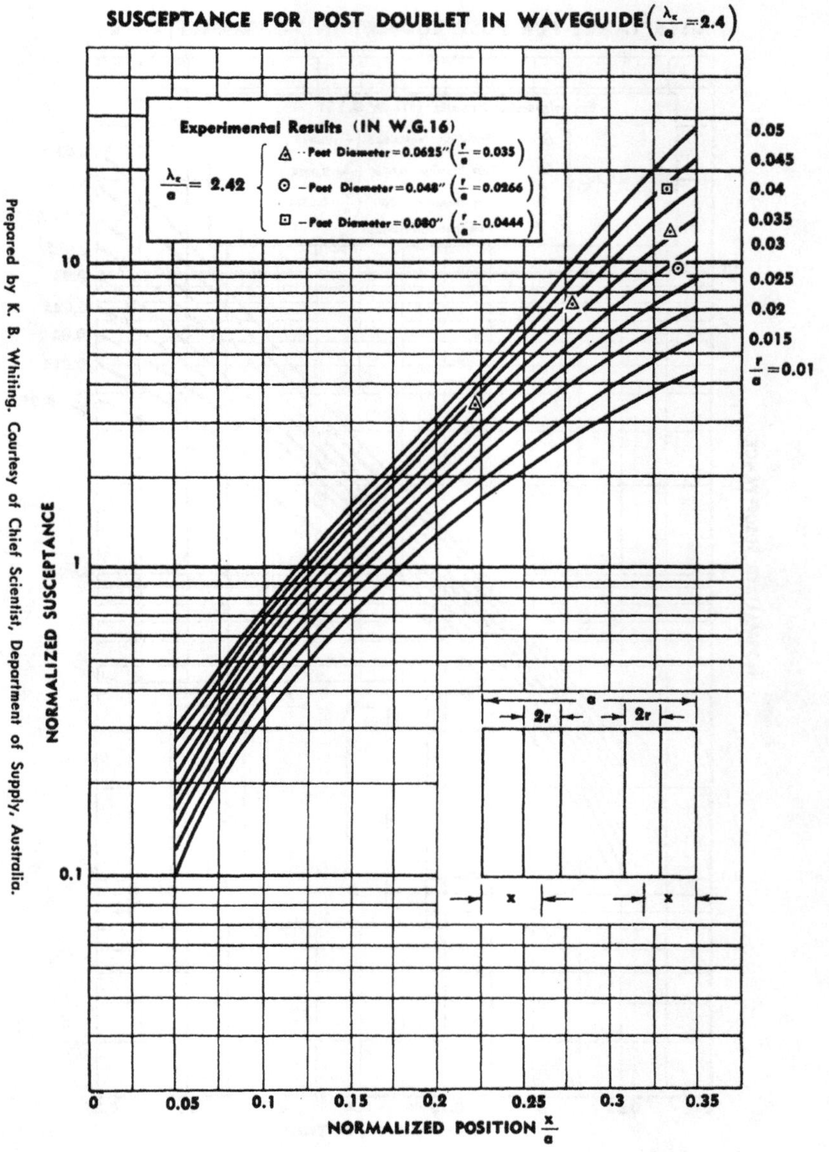

Figure 8.7(d) Susceptance for inductive post doublet in rectangular waveguide when $\lambda_g/a = 2.4$. (From Saad [10].)

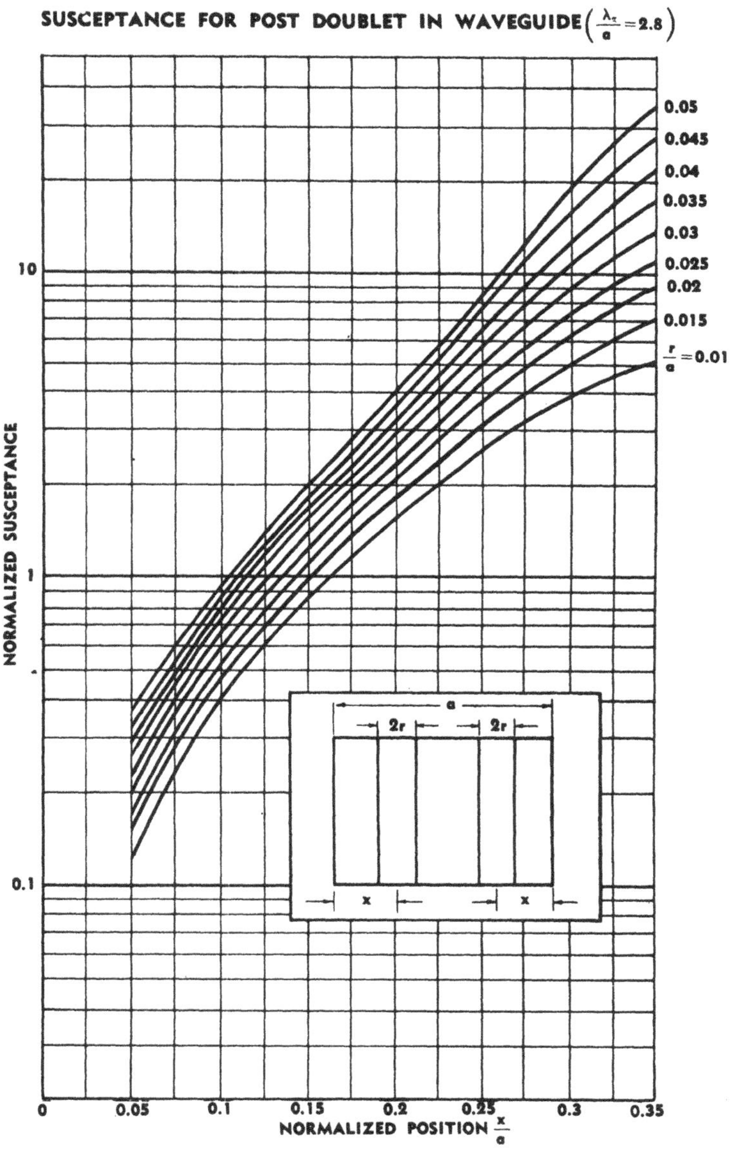

Figure 8.7(e) Susceptance for inductive post doublet in rectangular waveguide when $\lambda_g/a = 2.8$. (From Saad [10].)

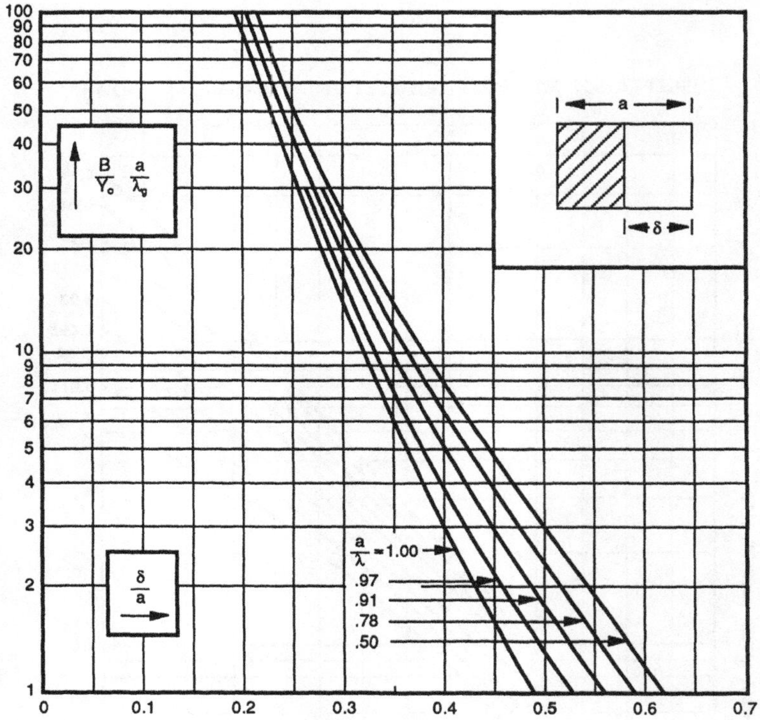

Figure 8.8(a) Susceptance of asymmetrical inductive thin iris in rectangular waveguide [see also Figure 8.8(b)]. (From Saad [10].)

We arbitrarily select a post with a diameter of 1/16 in. Then, $2r$ = 1/16 in. and $r = 0.0313$ in. Also, $r/a = 0.035$ and $\lambda_g/a = 1.74$. From Figure 8.6(b) for $\lambda_g/a = 1.6$, we determine that $d/a = 0.20$ and $d = 0.180$ in. From Figure 8.6(c) for $\lambda_g/a = 2.0$, we determine that $d/a = 0.16$ and $d = 0.144$ in. A linear interpolation between these values of d yields

$$d = 0.180 + \frac{1.74 - 1.60}{2.00 - 1.60}(0.144 - 0.180) = 0.167 \text{ in.}$$

To summarize, a 1/16-in. inductive post located 0.317 in. toward the load (from the position where the short had been located) and 0.167 in. from the side wall will match the load (VSWR $\approx$ 1.0).

This sample problem involved only one frequency; it was selected to demonstrate some of the "mechanics" of using a Smith chart and susceptance curves. In a real design problem, we probably would plot points for three to five frequencies across the band of interest.

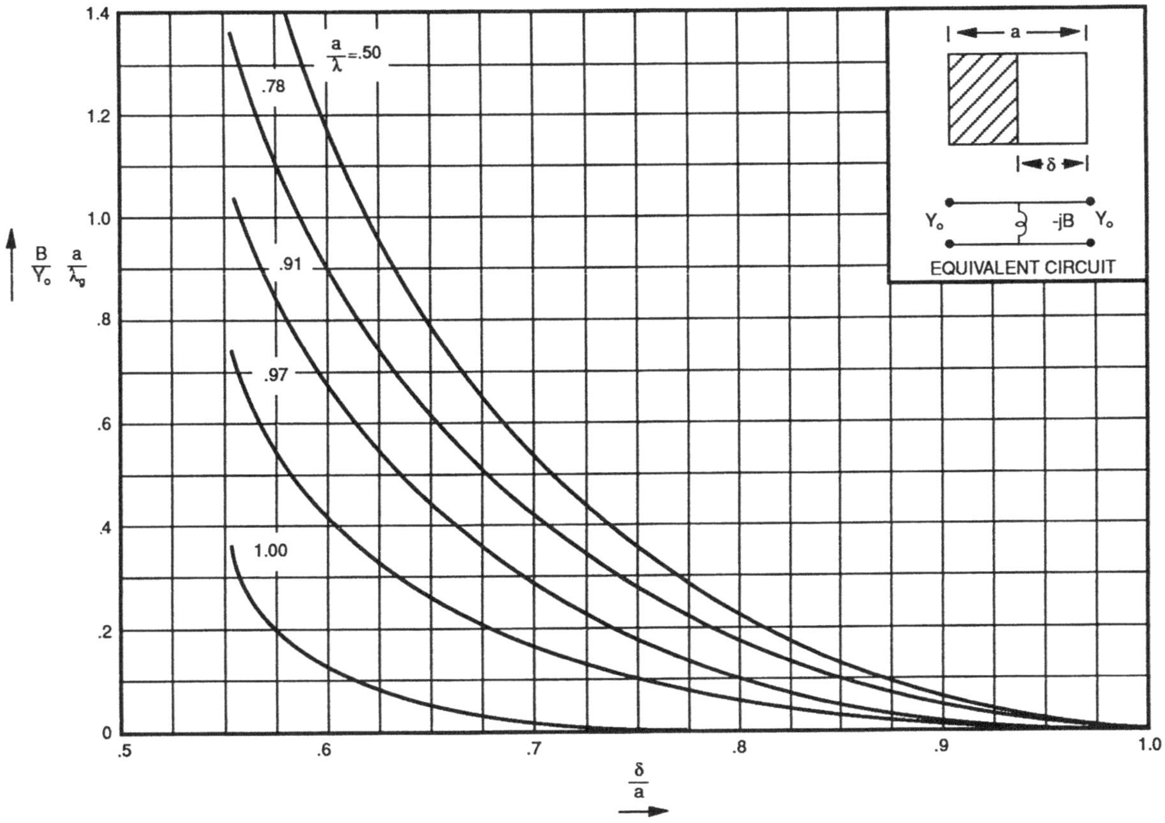

Figure 8.8(b) Susceptance of asymmetrical inductive thin iris in rectangular waveguide [see also Figure 8.8(a)]. (From Saad [10].)

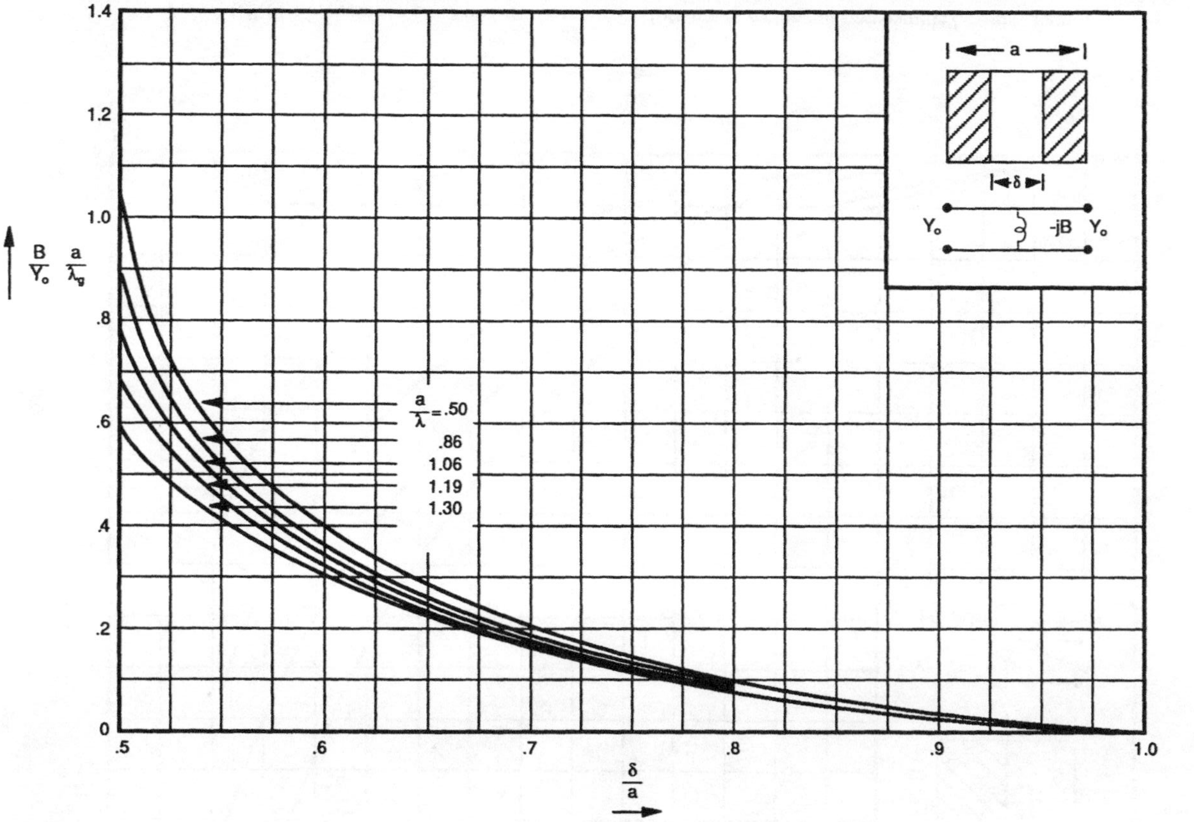

Figure 8.9(a) Susceptance of two symmetrical inductive thin irises in rectangular waveguide [see also Figure 8.9(b)]. (From Saad [10].)

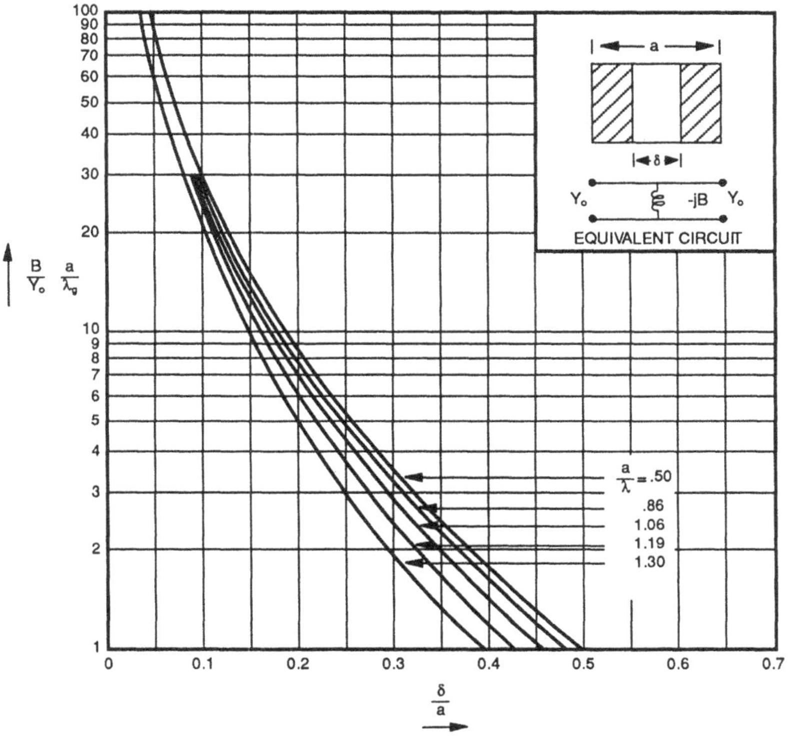

Figure 8.9(b) Susceptance of two symmetrical inductive thin irises in rectangular waveguide [see also Figure 8.9(a)]. (From Saad [10].)

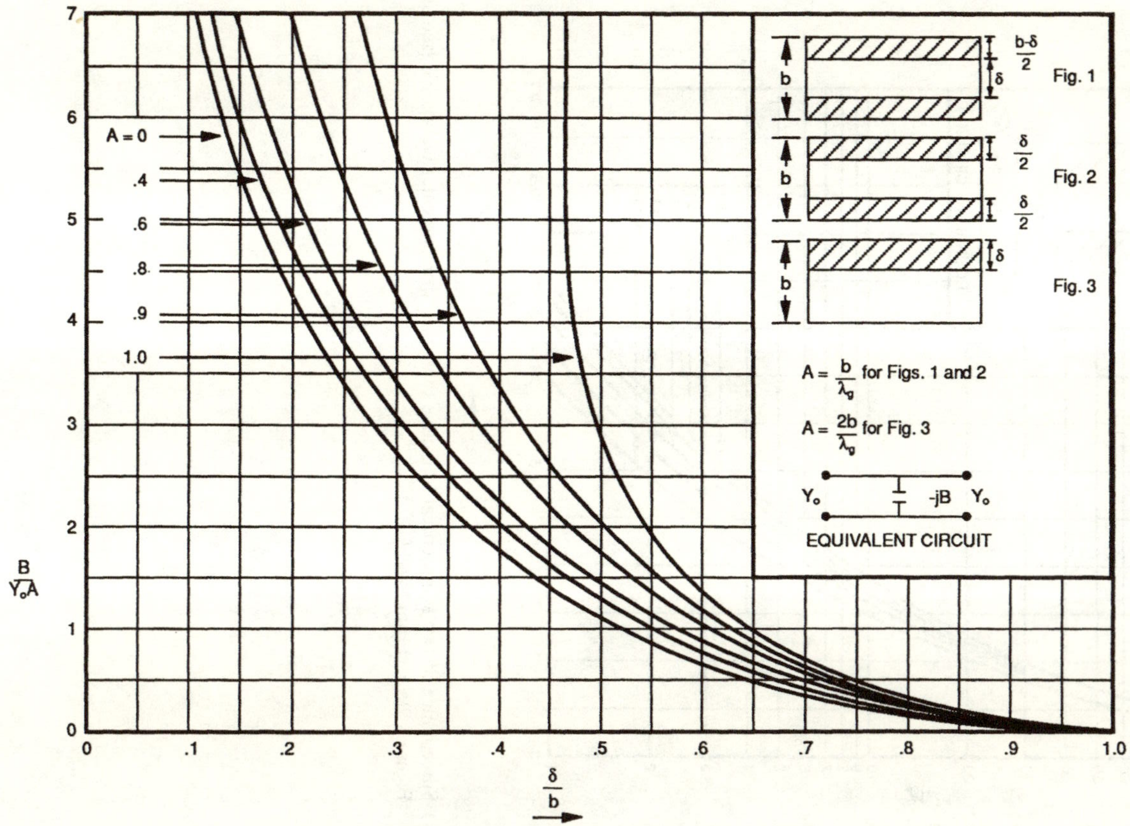

Figure 8.10 Susceptance of capacitive thin irises in rectangular waveguide. (From Saad [10].)

REFERENCES

1. G.L. Ragan, *Microwave Transmission Circuits,* MIT Radiation Laboratory Series, Vol. 9, New York: McGraw-Hill, 1948.
2. N. Marcuvitz, *Waveguide Handbook,* MIT Radiation Laboratory Series, Vol. 10, New York: McGraw-Hill, 1951.
3. T. Moreno, *Microwave Transmission Design Data,* New York: Dover Publications, 1958.
4. R.L. Thomas, *A Practical Introduction to Impedance Matching,* Norwood, MA: Artech House, 1976.
5. D.F. Bowman, "Impedance Matching and Broadbanding," Chapter 43 in *Antenna Engineering Handbook,* 2nd Ed. (R.C. Johnson and H. Jasik, eds.), New York: McGraw-Hill, 1984.
6. Y.C. Shih and T. Itoh, "Transmission Lines and Waveguides," Chapter 28 in *Antenna Handbook* (Y.T. Lo and S.W. Lee, eds.), New York: Van Nostrand Reinhold, 1988.
7. P.H. Smith, "Transmission Line Calculator," *Electron.,* Vol. 12, 1939, pp. 29–31.
8. P.H. Smith, "An Improved Transmission Line Calculator," *Electron.,* Vol. 17, 1944, pp. 130–133, 318–325.
9. S.Y. Liao, "Transmission Lines and Microwave Waveguides," Chapter 3 in *Microwave Devices and Circuits,* Englewood Cliffs, NJ: Prentice-Hall, 1980.
10. T.S. Saad, *Microwave Engineer's Handbook,* Vol. 1, Norwood, MA: Artech House, 1971.

LIST OF SYMBOLS

a	inside width of rectangular waveguide
B	susceptance
b	inside height of rectangular waveguide
G	conductance
d	distance of an inductive post from the inside wall of a rectangular waveguide
j	imaginary symbol, $j = \sqrt{-1}$
L_r, L_t	return loss and transmission loss, dB
P_{mo}	maximum power transmission on a matched line
P_{ms}	maximum power transmission on a line having a standing wave
r	radius of an inductive post
R	resistance
V_i, V_r, V_t	voltage (magnitudes) of incident, reflected, and transmitted waves, respectively
VSWR	voltage standing wave ratio
X	reactance
x	distance of post doublets from the inside walls of a rectangular waveguide
Y	admittance
Y_0	characteristic admittance of transmission line
Z	impedance
Z_0	characteristic impedance of transmission line

Z_L	load impedance
Γ	voltage reflection coefficient
δ	width or height of opening in waveguide iris
θ	phase angle of immittance phasor
λ_g	guide wavelength

Chapter 9
ANTENNA MEASUREMENTS

The ability to make valid antenna measurements requires knowledge in a broad field of technology. Fortunately, several excellent reference sources on this subject are available in the open literature [1–6]. This chapter will cover only a few special topics.

9.1 FAR-FIELD PHASE TAPER

Microwave antenna engineers usually divide the space surrounding an antenna into three regions: (1) the reactive near-field region, which dominates to a range of only about $\lambda/2\pi$, (2) the radiating near-field region, and (3) the radiating far-field region. In the radiating far-field region, the radiation pattern of the antenna under test (AUT) is essentially independent of the range (distance) to the source antenna. Thus, most antenna measurements (e.g., radiation patterns, gain, *et cetera*) should be made in the radiating far-field region.

One of the best known rules of thumb is that the far field begins at a range R such that

$$R = 2D^2/\lambda \tag{9.1}$$

where D is the maximum aperture dimension and λ is the wavelength. This rule of thumb was reasonable in the "old days" when most antennas had sidelobes of about 18 to 25 dB, but it is not adequate for pattern measurements when the AUT has very low sidelobes or for gain measurements when the AUT has a phase deviation across its aperture (e.g., a pyramidal horn). Well then, where does the far field begin? The answer is "it depends." For pattern measurements, it depends on the sidelobe level of the AUT and on the amount of acceptable measurement error.

The incident wave from a distant source antenna has a spherical shape centered at the source. This produces a phase taper across the aperture of the AUT as illustrated in Figure 9.1. If $R \gg \Delta R$, then (see Section A.16 in the Appendix):

$$\Delta R \approx \frac{D^2}{8R} \tag{9.2}$$

The phase taper at the edges of the AUT is

$$\Delta\phi \approx \frac{\pi}{4}\frac{D^2/\lambda}{R} \text{ radians} = 45\,\frac{D^2/\lambda}{R} \text{ degrees} \tag{9.3}$$

At $R = 2D^2/\lambda$, the phase taper at the edges is 22.5 degrees.

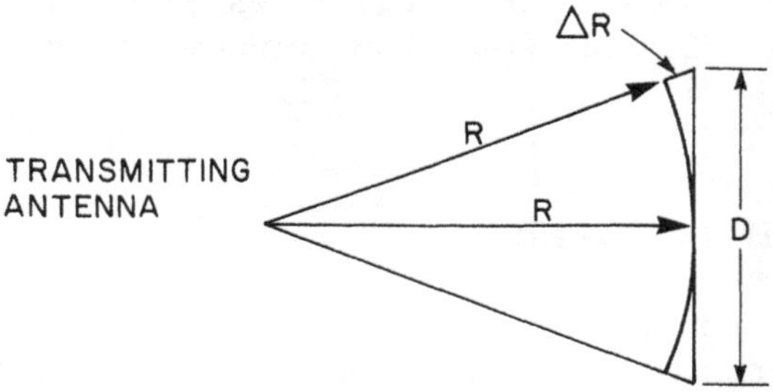

Figure 9.1 Illustration of phase taper across aperture D of a test antenna due to source transmitting from a finite distance R.

Measured antenna patterns behave essentially as follows. As the range to the source antenna decreases, the first null and the peak of the first sidelobe rise. The first sidelobe eventually becomes a shoulder on the main lobe and then it merges into the main lobe. As the range continues to decrease, the second sidelobe starts the same process of rising and merging. A point to remember is that an inadequate range to the source usually affects only the first couple of sidelobes.

Hansen [7] studied this situation theoretically using a line-source AUT with a Taylor $\bar{n}$ amplitude distribution. (Hansen selected the largest value of $\bar{n}$ that yields a monotonic aperture distribution.) He presented the results in a convenient graph as indicated in Figure 9.2. Note that if we have an AUT with −20-dB sidelobes and if we are willing to accept a 1.0-dB measurement error on the sidelobes, the required range to the source antenna is $2D^2/\lambda$. On the other hand, if the AUT has

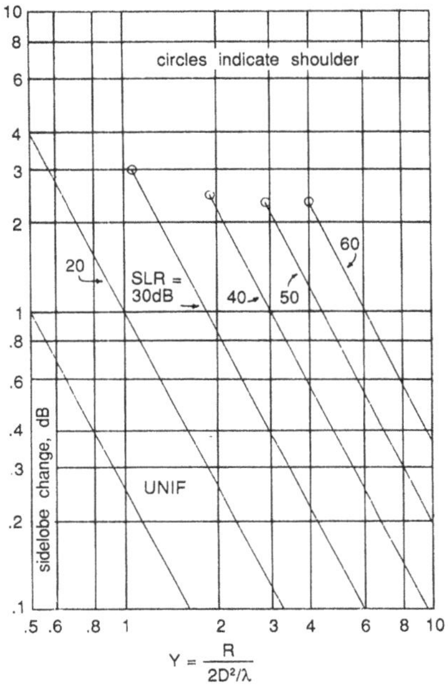

Figure 9.2 Sidelobe change *versus* normalized measurement distance for a Taylor $\bar{n}$ line source. (©1984 IEEE, from Hansen [7].)

−40-dB sidelobes and we will accept only a 0.6-dB measurement error, the required minimum range to the source is almost $8D^2/\lambda$.

Hansen indicates that the data in Figure 9.2 are universal curves that can be used for all linear antennas and for one dimension of a rectangular aperture antenna. The curves are easy to use and they enable us to choose quickly a suitable range for far-field measurements.

The phase taper also can reduce the measured gain, but the gain reduction [8] will be less than about 0.1 dB if $R \geq 2D^2/\lambda$.

9.2 FAR-FIELD AMPLITUDE TAPER

In an elevated antenna range, a high-gain source antenna is advantageous to reduce the likelihood that its narrow beam will illuminate the ground or other structures on the range. On the other hand, if the source beamwidth is too narrow, it produces an amplitude taper across the AUT. Such an amplitude taper will cause measurement errors in the gain and in the sidelobe levels.

Hansen [9] studied such situations using his circular aperture distributions [10], and he presented the results in two interesting graphs. Figure 9.3 illustrates sidelobe level errors *versus* amplitude taper for three AUTs with the indicated actual sidelobe levels. The measured sidelobe levels will be less than the actual levels. Figure 9.4 illustrates the gain measurement error *versus* amplitude taper for the same three AUTs. The measured gains will be less than the actual gains. Note that the largest error occurs with the 25-dB sidelobe case; this is because of the relative rates of change of beamwidth and of sidelobe level with Hansen's parameter H. Hansen points out that the directivity-beamwidth squared product peaks at a sidelobe level of roughly -30 dB.

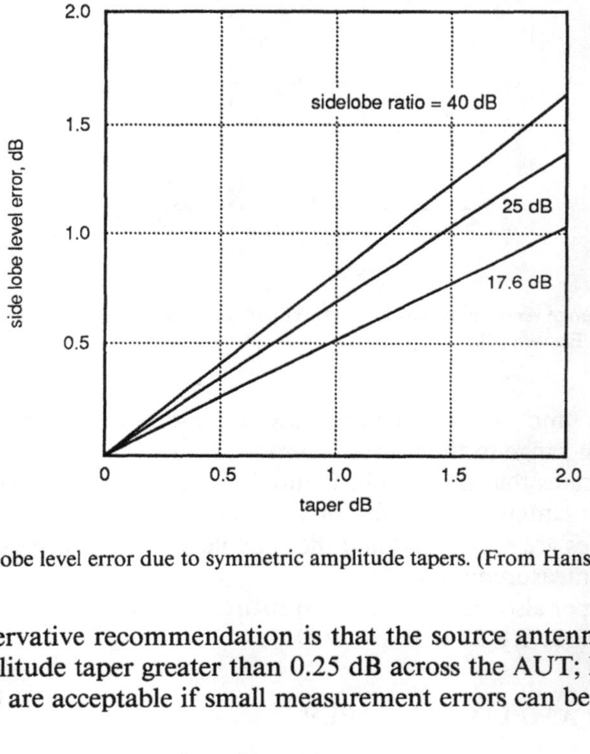

Figure 9.3 Sidelobe level error due to symmetric amplitude tapers. (From Hansen [9].)

A conservative recommendation is that the source antenna should not produce an amplitude taper greater than 0.25 dB across the AUT; however, tapers of 0.5 or 1.0 dB are acceptable if small measurement errors can be tolerated.

9.3 EXTRANEOUS COHERENT SIGNALS

For accurate antenna measurements, we want a single plane wave incident on the AUT; however, in essentially all cases, undesired reflections will occur that introduce extraneous coherent waves incident on the AUT. For the case of a single extraneous wave, assume that at the terminals of the AUT the direct-path wave produces

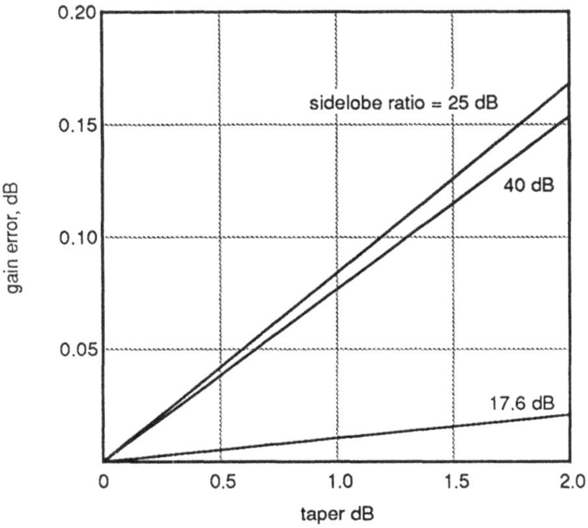

Figure 9.4 Gain error due to symmetric amplitude tapers. (From Hansen [9].)

a voltage E_d and the extraneous wave produces a voltage E_x. For typical logarithmic pattern recordings (Chap. 14 [2]), the error in measured levels can be

$$\Delta L = 20 \log \left(\frac{E_d \pm E_x}{E_d} \right) \quad \text{dB} \tag{9.4}$$

for $E_d > E_x$. In the less likely event that $E_d < E_x$,

$$\Delta L = 20 \log \left(\frac{E_x \pm E_d}{E_d} \right) \quad \text{dB} \tag{9.5}$$

Results of calculations using these equations are illustrated in Figure 9.5.

9.4 POWER TRANSFER

Consider power that is transferred from a source antenna to the AUT; assume that the antennas are in free space and are separated by a large distance R (in the far field of each other). The received power is

$$P_r = \frac{G_t P_t}{4\pi R^2} \frac{G_r \lambda^2}{4\pi} = \left(\frac{\lambda}{4\pi R} \right)^2 G_t G_r P_t \tag{9.6}$$

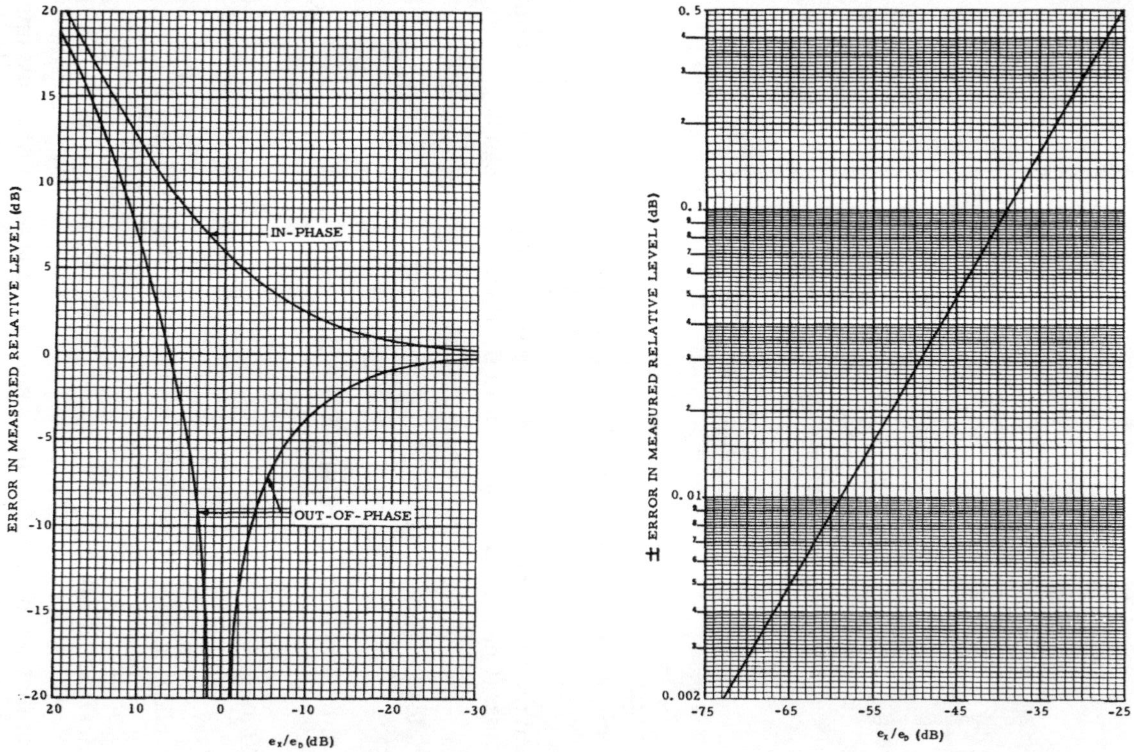

Figure 9.5 Maximum possible error in measured relative pattern level due to a coherent extraneous signal. Linear scales are used for signal ratios of +20 to −30 dB; the ± errors are essentially equal for ratios of −25 dB or less and are indicated in the logarithmic plot for ratios down to −75 dB. (From Hollis *et al.* [2].)

where

P_r, P_t = received power and transmitted power, respectively
G_r, G_t = gain of receiving antenna (AUT) and gain of transmitting antenna (source), respectively
λ = wavelength
R = range (distance) between antennas

This equation is a form of the Friis transmission formula. In logarithmic form,

$$L_r = L_t + g_t + g_r - 20\log(4\pi R/\lambda) \tag{9.7}$$

where

L_r = signal level at the output terminals of the receiving antenna (AUT), dBm
L_t = signal level at the input terminals of the transmitting antenna (source), dBm
g_t = $10\log G_t$
g_r = $10\log G_r$

Note that the gains are the values of each antenna in the direction toward the other antenna, not necessarily the peak gains of the main lobes.

The term

$$N = 20\log(4\pi R/\lambda) \tag{9.8}$$

is called *space attenuation,* and it can be evaluated quickly using the nomograph in Figure 9.6.

If we are making radar cross-section measurements, the received power is

$$P_r = \frac{G_t P_t}{4\pi R^2}\frac{\sigma}{4\pi R^2}\frac{G_r\lambda^2}{4\pi} \tag{9.9}$$

where σ is the radar cross section. If the same antenna is used for both transmission and reception of energy, then $G_t = G_r = G$, and the received power may be written as

$$P_r = \frac{G^2\lambda^2\sigma}{(4\pi)^3R^4}P_t \tag{9.10}$$

which is a simple form of the radar equation.

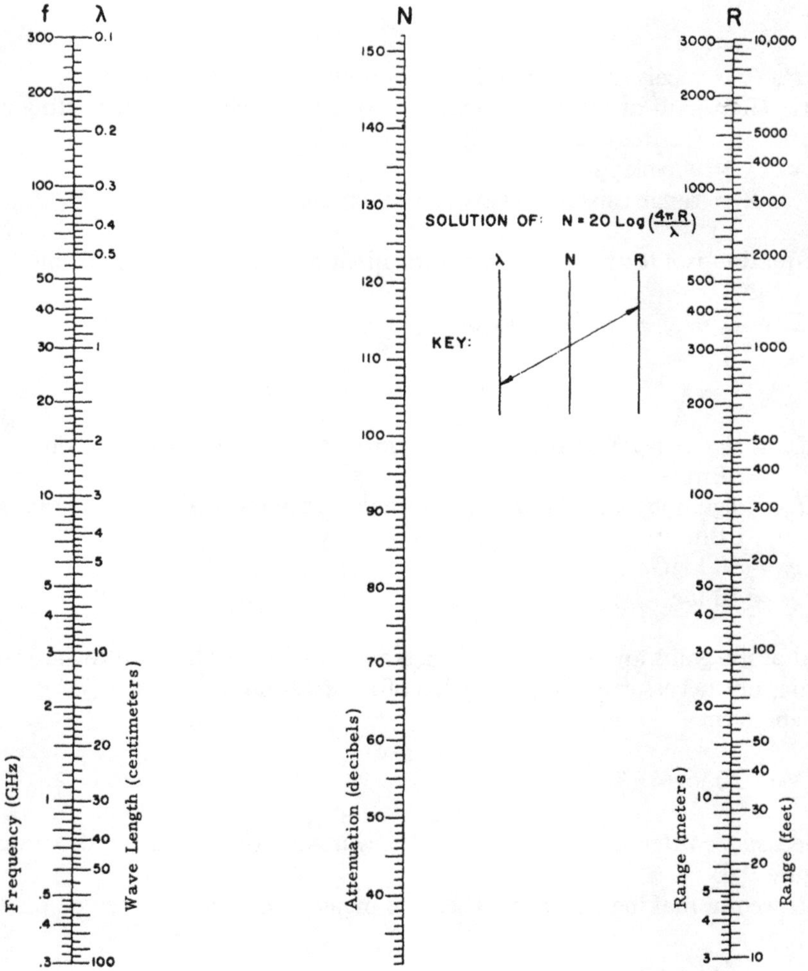

Figure 9.6 Nomograph of space attenuation as a function of wavelength and range. If the nomograph does not cover the desired range of λ and *R*, multiply both scales by *n* and do not change the *N* scale. (From Hollis *et al.* [2].)

9.5 FOCUSING TECHNIQUES

Earlier in this chapter, we discussed far-zone range requirements to reduce the phase taper across the aperture of the AUT due to the spherical wavefront from the source antenna. This phase taper can be reduced greatly by focusing the AUT at the range (distance) to the source antenna. Such focusing techniques apply to paraboloids and lenses that can be focused by axial positioning of the feed, to arrays which can be bent to conform to the spherical wavefront, and to electronically phased arrays, the element phasing of which can be altered to provide focusing at the test distance. Reference [11] presents a review of focusing techniques.

The geometry relating to focusing techniques is presented in Figure 9.7. When the phase center of the primary feed is located at F, the focal point of the reflector, the antenna is focused at infinity. As the feed is moved axially away from the reflector through a distance ϵ to F', the reflected waves tend to focus at some point F'' on the paraboloid's axis. Some aberrations now must exist because an ellipsoidal reflector is required to provide aberration-free focusing at a finite range.

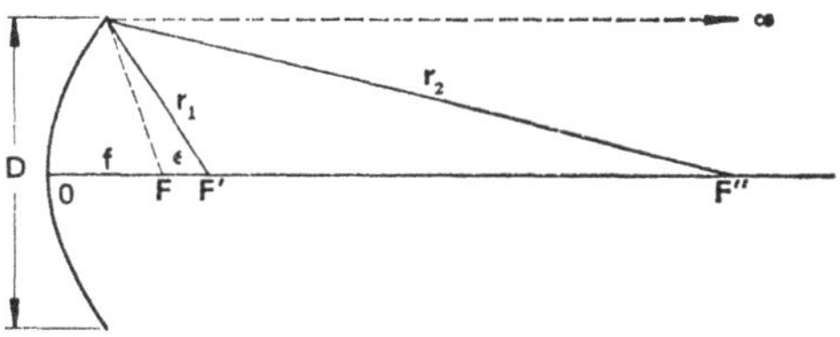

Figure 9.7 Focusing geometry for a paraboloidal reflector. (©1973 IEEE, from Johnson *et al.* [11].)

For a given reflector diameter D, the aberration phase errors increase as OF'' decreases, as the focal length F decreases, and as the frequency increases; however, a useful range exists through which the phase errors are tolerable.

Cheng and Moseley [12] proposed the following as a criterion for focusing the condition:

$$r_1 + r_2 = \epsilon + OF + OF'' \tag{9.11}$$

where the terms are illustrated in Figure 9.7. This criterion for focusing leads to

$$\epsilon \approx \frac{1}{R}\left[F^2 + \left(\frac{D}{4}\right)^2\right] \tag{9.12}$$

Many patterns were calculated by Fourier transformation of the aperture field for separations from $R = D^2/16\lambda$ to ∞ and for various axial feed positions [11]. The feed position was given by $k\epsilon$ where k varied between zero and unity. The deepest nulls occurred for values of k between 0.92 and 0.93 for all separations.

Figure 9.8 is a set of calculated patterns for an antenna, having a 0.375 F/D ratio, focused and tested from $R = 2D^2/\lambda$ down to $D^2/16\lambda$. The pattern at $R =$

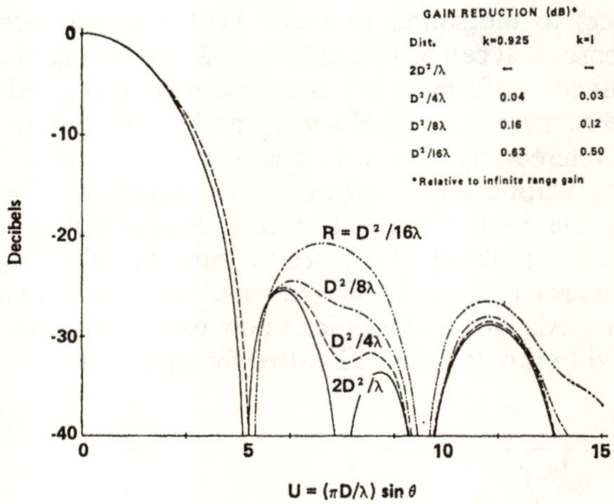

Figure 9.8 Calculated principal-plane patterns of a paraboloid focused at the indicated test separations. The pattern for $2D^2/\lambda$ is virtually identical to the infinite separation pattern. The deepest nulls occur for k near 0.925. (©1973 IEEE, from Johnson *et al.* [11].)

$2D^2/\lambda$ is virtually identical with the infinite separation range pattern. The sidelobes are seen to deviate from those in the $2D^2/\lambda$ pattern as the range gets shorter. Other patterns calculated for an F/D ratio of 0.25 show similar character except that a given distortion occurs at slightly less than twice the separation that it occurs at for the 0.375 F/D ratio.

Antenna focusing is a very useful technique, but it should be used with caution. Suggestions have been made that for reflectors having F/D ratios of about 0.35 or greater, a range as short as $D^2/2\lambda$ is acceptable unless extreme accuracy is required. The AUT should be focused for deepest nulls, and then gain and patterns (main lobe and the first couple of sidelobes) can be measured. To refocus the antenna to infinity, the feed should be moved 0.925ϵ toward the reflector. For shorter ranges, one should become more familiar with the literature before proceeding.

With Cassegrain antennas, focusing should be accomplished through movement of the subreflector because the focusing effects are not very sensitive to feed movements. Figure 9.9 illustrates some pertinent parameters for Cassegrain antennas. A parameter known as the *magnification* can be calculated from

$$M = \frac{D}{4F_p} \cot \frac{\phi_e}{2} \tag{9.13}$$

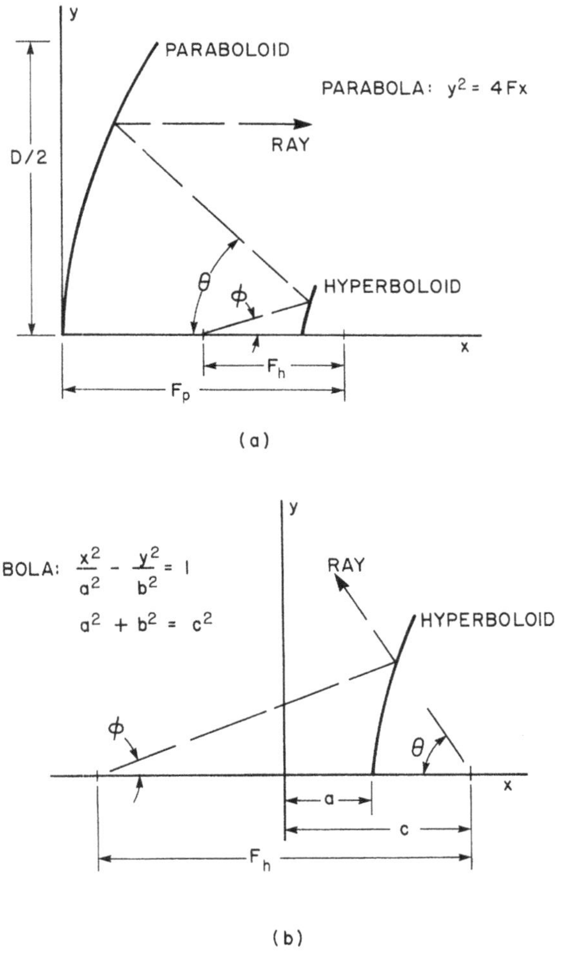

Figure 9.9 Side view of the upper half of a Cassegrain antenna: (a) both reflectors; (b) focal region in a shifted coordinate system.

where

M = magnification
D = diameter of paraboloid (main reflector)
F_p = focal length of paraboloid
ϕ_e = angle of edge ray relative to antenna axis as seen from the feed

For Cassegrain antennas having magnifications of about 4 or greater, satisfactory focusing results can be obtained by moving the subreflector the same distance as the feed for a front-fed reflector having the same diameter and focal length as the main reflector.

A more accurate estimate, particularly for antennas with low magnification, of the required subreflector movement for focusing can be derived as follows. Figure 9.10(a) depicts a front-fed paraboloid in which the feed is moved a distance ϵ. If $\epsilon \ll r_1$, then $\theta \approx \theta'$, and the region around the focal point is as shown in Figure 9.10(b). The path length of the central ray is increased by a distance ϵ, and the path lengths of other rays are increased by a distance $\epsilon \cos\theta$. Thus, the relative path length change is

$$\Delta_{ff} \approx \epsilon_{ff}(1 - \cos\theta) \tag{9.14}$$

where the *ff* subscripts indicate that the parameters are for movement of the feed of a front-fed paraboloid.

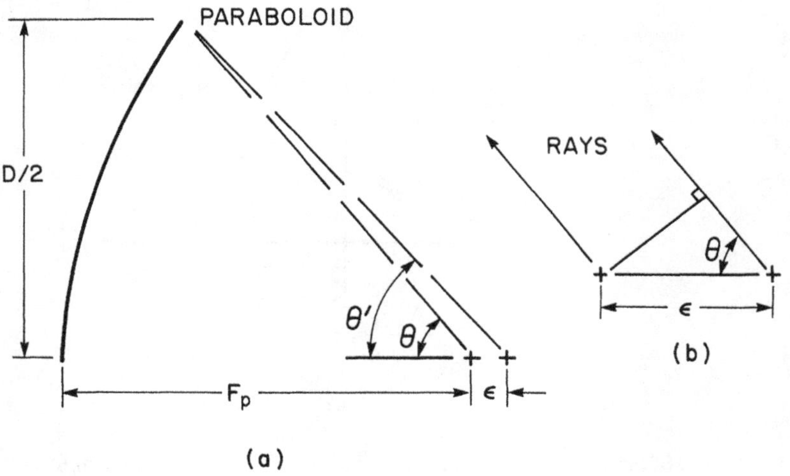

Figure 9.10 Side view of the upper half of a front-fed paraboloidal reflector that is focused by feed movement: (a) reflector and feed; (b) focal region.

For a Cassegrain antenna, assume that the region around the reflection point on the subreflector is a planar surface and that the movement ϵ is small compared to the major dimensions of the antenna. The region around the reflection point is depicted in Figure 9.11; note that the curved subreflector appears to be tilted by an

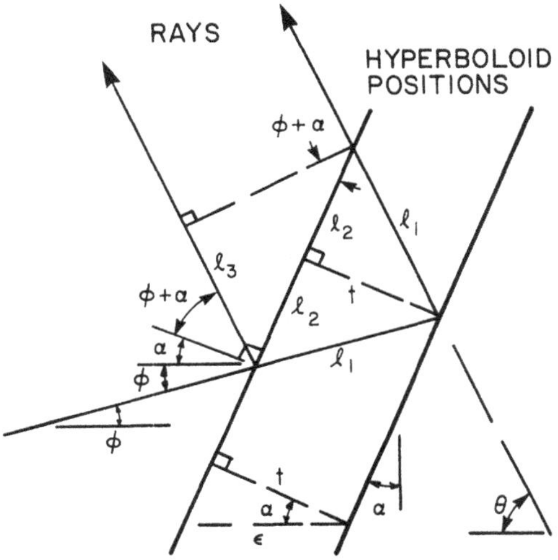

Figure 9.11 Reflection from a hyperboloidal subreflector that has been moved axially a distance ϵ.

angle α at the reflection point. The path length of the central ray is increased by a distance 2ϵ, and the path lengths of other rays are increased by a distance $2l_1 - l_3$. The relative path-length change is

$$\Delta_{cg} \approx 2\epsilon_{cg}\left[1 - \cos\left(\frac{\theta - \phi}{2}\right)\cos\left(\frac{\theta + \phi}{2}\right)\right] \tag{9.15}$$

where the *cg* subscripts indicate that the parameters are for the movement of the subreflector of a Cassegrain antenna.

If we equate the two relative path-length changes, we have

$$\epsilon_{cg} \approx \frac{1 - \cos\theta}{2\left[1 - \cos\left(\frac{\theta - \phi}{2}\right)\cos\left(\frac{\theta + \phi}{2}\right)\right]}\,\epsilon_{ff} \tag{9.16}$$

In other words, this equation tells us how far to move the subreflector to produce essentially the same focusing effect as the movement of a front feed.

We now write eq. (9.12) as

$$\epsilon_{ff} \approx \frac{1}{R}\left[F_p^2 + \left(\frac{D}{4}\right)^2\right] \tag{9.17}$$

to indicate that it is for a front-fed paraboloidal reflector.

A typical measurement problem in the radiating near-field region of a Cassegrain antenna could then proceed as follows:

1. Focus the antenna experimentally (usually for the deepest first nulls) by moving the subreflector axially on the selected antenna test range of length R.
2. Measure gain and patterns (through the first couple of sidelobes); these data represent far-field characteristics.
3. Assume that the main reflector is operated as a front-fed paraboloid, and calculate the required feed movement ϵ_{ff} to focus at the range R using eq. (9.17).
4. Calculate the required subreflector movement ϵ_{cg} to focus the Cassegrain antenna at the range R using the angle values for the edge ray in eq. (9.16).
5. Move the subreflector axially toward the main reflector a distance 0.925 ϵ_{cg} to focus the antenna to infinity.
6. Measure wide-angle characteristics as desired.

9.6 PRIMARY PATTERNS

The measurement of broad-beam primary patterns seems to be straightforward, but a simple mistake is made quite often. To make accurate pattern measurements, the phase center of the feed (AUT) must be on the axis of rotation of the arm that moves the probe in a circular path.

Figure 9.12 illustrates a case in which the phase center of the feed is located a distance δ forward of the axis of rotation of the arm that moves the probe antenna. We assume that the probe is at the pattern angle θ and at a distance R from the feed; however, the pattern angle is θ' and the distance is R'.

When the probe is on the feed axis, both θ and θ' equal zero and no angle error occurs, but when the probe is at $\theta = 90$ degrees, the actual pattern angle is $\theta' = 90 + 180\delta/R\pi$ degrees. For example, if $\delta = 2$ in. and $R = 40$ in., then $\theta' = 92.9$ degrees—an error of 2.9 degrees. Also, when $\theta = 0$ degrees, the separation is $R - \delta$; but when $\theta = 90$ degrees, the separation is R. In the same example, the relative space attenuation (error) is $20 \log R/(R - \delta) = 0.45$ dB. If phase-measuring equipment is not available for properly positioning the feed, a theoretical estimate of the location of the phase center should be used.

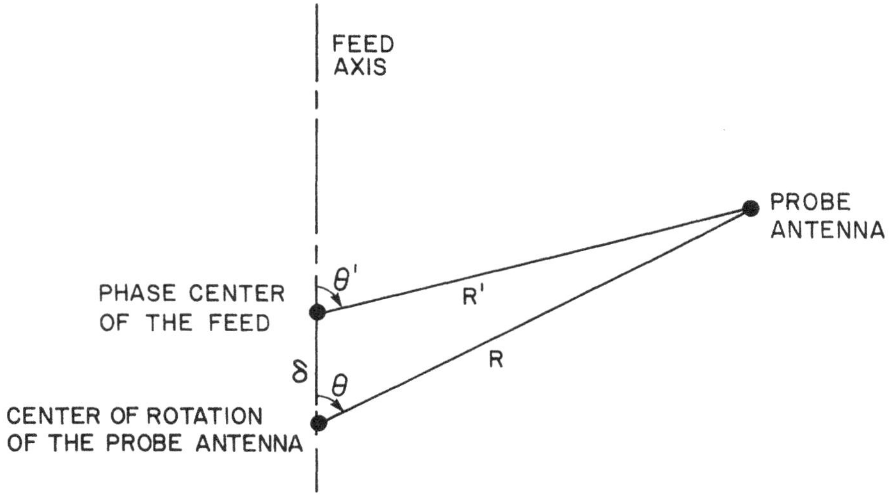

Figure 9.12 Illustration of a case in which the phase center of the feed is located a distance δ forward of the axis of rotation of the arm that moves the probe antenna.

REFERENCES

1. *IEEE Standard Test Procedures for Antennas,* IEEE Standard 149-1979, New York: Institute of Electrical and Electronics Engineers, 1979.
2. J.S. Hollis, T.J. Lyon, and L. Clayton, *Microwave Antenna Measurements,* Atlanta, GA: Scientific-Atlanta, 1970.
3. C.C. Cutler, A.P. King, and W.E. Kock, "Microwave Antenna Measurements," *Proc. IRE,* Vol. 35, December 1947, pp. 1462–1471.
4. W.H. Kummer and E.S. Gillespie, "Antenna Measurements—1978," *Proc. IEEE,* Vol. 66, April 1978, pp. 483–507.
5. J. Appel-Hansen, J.D. Dyson, E.S. Gillespie, and T.G. Hickman, "Antenna Measurements," Chapter 8 in *The Handbook of Antenna Design* (A.W. Rudge, K. Milne, A.D. Olver, and P. Knight, eds.), Vol. 1, London: Peter Peregrinus, 1982.
6. E.S. Gillespie, "Measurement of Antenna Radiation Characteristics on Far-Field Ranges," Chapter 32 in *Antenna Handbook* (Y.T. Lo and S.W. Lee, eds.), New York: Van Nostrand Reinhold, 1988.
7. R.C. Hansen, "Measurement Distance Effects on Low Sidelobe Patterns," *IEEE Trans. Antennas Propagat.,* Vol. AP-32, June 1984, pp. 591–594.
8. P.S. Hacker and H.E. Schrank, "Range Distance Requirements for Measuring Low and Ultralow Sidelobe Antenna Patterns," *IEEE Trans. Antennas Propagat.,* Vol. AP-30, September 1982, pp. 956–966.
9. R.C. Hansen, "Effect of Field Amplitude Taper on Measured Antenna Gain and Sidelobes," *Electron. Lett.,* Vol. 17, April 2, 1981, pp. 260–261.

10. R.C. Hansen, "A One-Parameter Circular Aperture Distribution with Narrow Beamwidth and Low Sidelobes," *IEEE Trans. Antennas Propagat.*, Vol. AP-24, July 1976, pp. 477–480.
11. R.C. Johnson, H.A. Ecker, and J.S. Hollis, "Determination of Far-Field Antenna Patterns from Near-Field Measurements," *Proc. IEEE,* Vol. 61, December 1973, pp. 1668–1694.
12. D.K. Cheng and S.T. Moseley, "On-Axis Defocus Characteristics of the Paraboloidal Reflector," *IRE Trans. Antennas Propagat.*, Vol. AP-3, October 1955, pp. 214–216.

LIST OF SYMBOLS

AUT	antenna under test
a, b, c	constants for defining hyperbola
D	maximum aperture dimension (diameter of main reflector)
E_d	received signal voltage by AUT due to the direct wave
E_x	received signal voltage by AUT due to the extraneous wave
F	focal length
F_h	focal length of hyperboloid
F_p	focal length of paraboloid
G	gain of common receiving and transmitting antenna
G_r	gain of receiving antenna (AUT)
G_t	gain of transmitting antenna (source)
g_r	$10 \log G_r$
g_t	$10 \log G_t$
k	arbitrary constant (value 0 to 1) multiplier of ϵ for feed movement for focusing
L_r	signal level at the output terminals of the receiving antenna (AUT), dBm
L_t	signal level at the input terminals of the source antenna (transmitting), dBm
l_n	(where $n = 1, 2,$ or 3) lengths of line segments
M	magnification of a Cassegrain antenna
N	space attenuation
OF, OF''	distances illustrated in Figure 9.7
P_r	received power
P_t	transmitted power
R	range (distance) from AUT to source antenna or probe
R'	actual range from feed phase center to probe
r_1, r_2	distances illustrated in Figure 9.7
t	normal separation between hyperboloidal subreflector positions before and after focusing
x, y	coordinates
α	tilt angle of hyperboloidal subreflector at the point of ray reflection

ΔR	difference in range from source antenna to center of AUT and to edge of AUT
δ	separation of feed phase center and center of rotation of probe on a primary pattern range
ΔL	error in measured power level due to an extraneous wave
$\Delta\phi$	phase error at the edges of the AUT
ϵ	distance of feed or subreflector movement for focusing
θ	angle of ray relative to antenna axis as seen from the focal point of the main reflector, and pattern angle indicated by probe position on a feed pattern range
θ'	angle of ray relative to antenna axis as seen from the focal point of the main reflector after focusing the feed, and actual pattern angle of a feed
λ	wavelength
σ	radar cross section
ϕ	angle of ray relative to antenna axis as seen from the feed of a Cassegrain antenna; a subscript e refers to the edge ray

Chapter 10
PATTERN "TALK"

Much of the data encountered by an antenna engineer are in the form of patterns, and often the engineer reads little more from the patterns than "does the antenna or component meet specifications or goals?" With experience, we realize that the pattern "wiggles" often say much more. In this chapter, some of the interesting methods for reading and interpreting various patterns will be presented.

10.1 VSWR PATTERNS

VSWR (or return-loss) *versus* frequency patterns are very common. In many cases, the VSWR periodically peaks due to constructive interference from two separated discontinuities. Figure 10.1 depicts a transmission line with two discontinuities separated by a distance L. Let

- L = distance between two reflecting discontinuities
- f_1, f_2 = the frequencies at two adjacent VSWR peaks
- $\lambda_{g1}, \lambda_{g2}$ = guide wavelengths corresponding to f_1 and f_2
- n = an integer

Adjacent VSWR peaks occur when

$$L = n\frac{\lambda_{g1}}{2} = (n + 1)\frac{\lambda_{g2}}{2} \tag{10.1}$$

Thus,

$$n = \frac{\lambda_{g2}}{\lambda_{g1} - \lambda_{g2}} \tag{10.2}$$

From these two equations,

$$L = \frac{\lambda_{g1}\lambda_{g2}}{2(\lambda_{g1} - \lambda_{g2})} \tag{10.3}$$

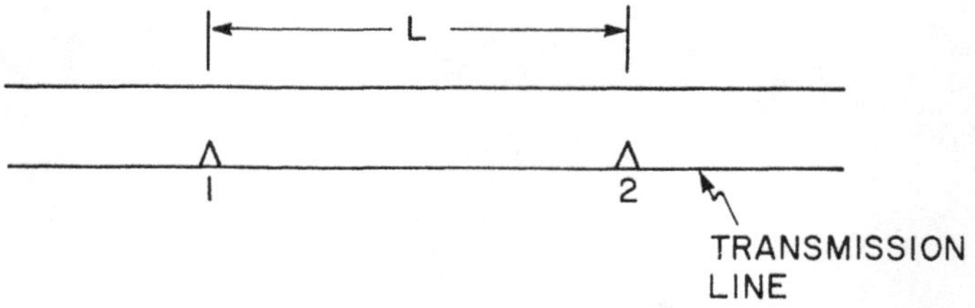

Figure 10.1 Two reflecting discontinuities, separated by a distance L, in a transmission line.

Note that in eq. (10.1) we assumed that $\lambda_{g1} > \lambda_{g2}$. The corresponding assumption for frequencies is that $f_1 < f_2$.

For example, suppose that in WR-90 waveguide we observe periodic VSWR peaks with two adjacent peaks at 10.0 and 10.3 GHz. These frequencies correspond to wavelengths of 1.563 and 1.486 in. From eq. (10.3), L = 15.1 in. Thus, we should look at the DUT (device under test) to locate two reflecting discontinuities that are about 15 in. apart. Then, we can work on these discontinuities to reduce the peak VSWR.

10.2 PRIMARY PATTERNS

In many cases, the recording of primary (broad-beamwidth) patterns without degradations from undesired stray reflections from objects in the test chamber is difficult; however, with some effort to locate and eliminate stray reflections, we can usually record accurate and "smooth" patterns. (Do not overlook the importance of designing the probe to help discriminate against stray reflections.)

Figure 10.2 depicts a simple two-dimensional model of a range for recording primary patterns. A broad-beamwidth feed is located with its phase center at point 1, and a probe moves along an arc with a radius R_1 centered at point 1. A stray isotropic reflecting source is located to the right side of the feed at a distance δ. Assume that the primary pattern is parabolic when plotted as power in dB *versus* pattern angle. Let

- θ = pattern angle
- BW_{10} = 10-dB beamwidth of primary pattern
- dB_θ = power level from the feed at pattern angle θ, dB
- dB_2 = power level of the radiation from the stray reflecting source relative to the radiation from the feed on axis, dB
- dB_t = resultant power level, dB

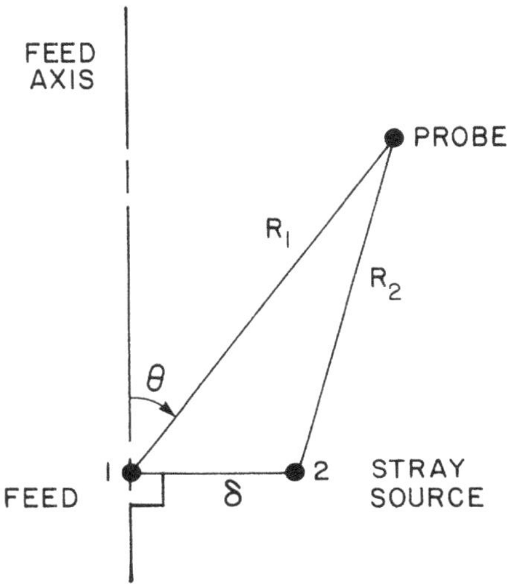

Figure 10.2 A simple two-dimensional model of a range for recording primary patterns. The feed is located at point 1, and a stray reflecting source is located at point 2.

- f = frequency
- R_1, R_2 = distances to probe from feed and from stray source
- δ = distance from feed to stray reflecting source
- E_1, E_2 = relative field strength at probe due to feed and due to stray reflecting source
- E_t = resultant relative field strength
- ψ = phase of E_2 relative to the phase of E_1
- ψ_c = arbitrary phase-shift constant
- λ = wavelength

From the parabolic beamshape,

$$\mathrm{dB}_\theta = -10\left(\frac{2\theta}{\mathrm{BW}_{10}}\right)^2 = 20\log E_1 \quad (10.4)$$

$$E_1 = 10^{-1/2(2\theta/\mathrm{BW}_{10})^2} \quad (10.5)$$

From the law of cosines,

$$R_2^2 = \delta^2 + R_1^2 - 2\delta R_1 \sin\theta \quad (10.6)$$

Also,

$$\psi = \frac{2\pi}{\lambda}[R_1 - (\delta + R_2)] + \psi_c \tag{10.7}$$

$$dB_2 = 20 \log E_2 \tag{10.8}$$

$$E_2 = 10^{(dB_2/20)} \tag{10.9}$$

The three field-strength phasors are related as indicated in Figure 10.3. Thus,

$$E_t^2 = E_1^2 + E_2^2 - 2E_1E_2 \cos(\pi - \psi) \tag{10.10}$$

$$dB_t = 20 \log E_t \tag{10.11}$$

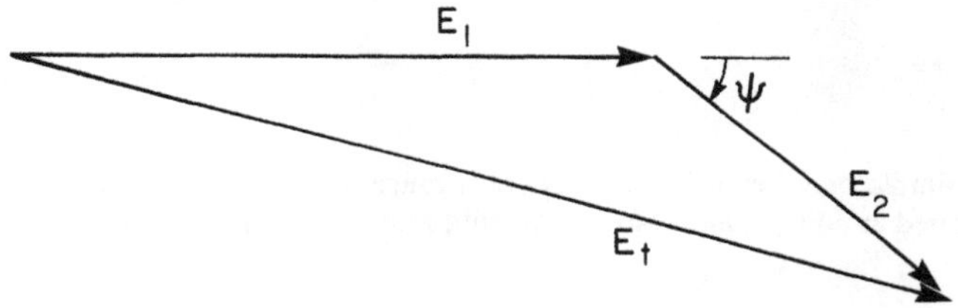

Figure 10.3 Relationship of three field-strength phasors.

The above two-dimensional model is rather simple when compared to the real world, but it can be used to indicate how a stray reflecting source can affect the recorded radiation pattern. For example, let f = 10 GHz, R_1 = 40 in., BW_{10} = 120 degrees, $dB_2 = -30$, and ψ_c = 0 degrees. Patterns were calculated for three locations of the stray reflection source: δ = 1, 10, and 30 in.; the results are shown in Figures 10.4 through 10.6. In each figure, the dash-dot line is the parabolic pattern with no stray reflections, and the solid line is the pattern that is perturbed by a stray source at a level of −30 dB relative to the feed on axis.

In Figure 10.4, the stray source is only 1 in. from the phase center of the feed. Note that the resultant pattern (solid line) is distorted; however, we might not notice the distortion without the unperturbed pattern (dash-dot line) shown for comparison. If the feed is symmetric, then the radiation pattern should be symmetric. The symmetry can be checked by recording a pattern, folding it along the center line, and viewing the two halves of the pattern on a light table. If the pattern is asymmetric, then an unknown asymmetry exists in the feed or a stray source

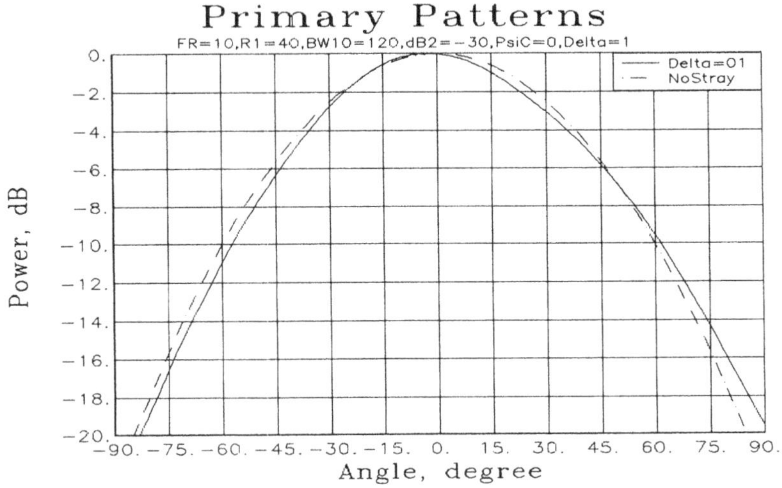

Figure 10.4 Patterns with and without a stray reflecting source located 1 in. from the feed.

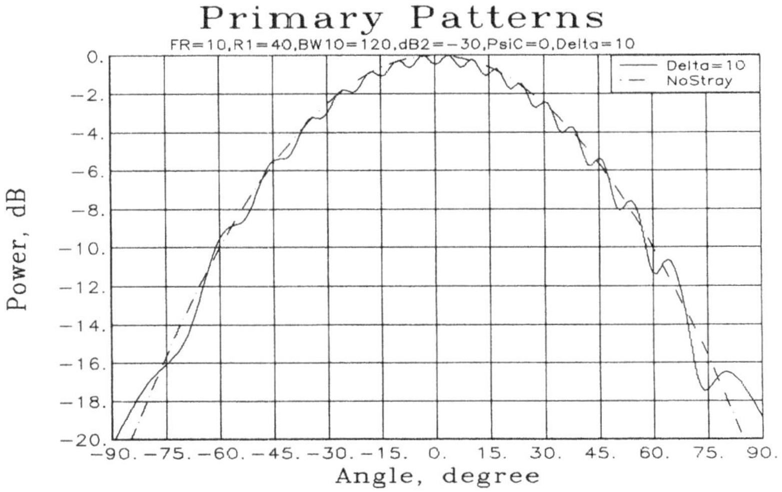

Figure 10.5 Patterns with and without a stray reflecting source located 10 in. from the feed.

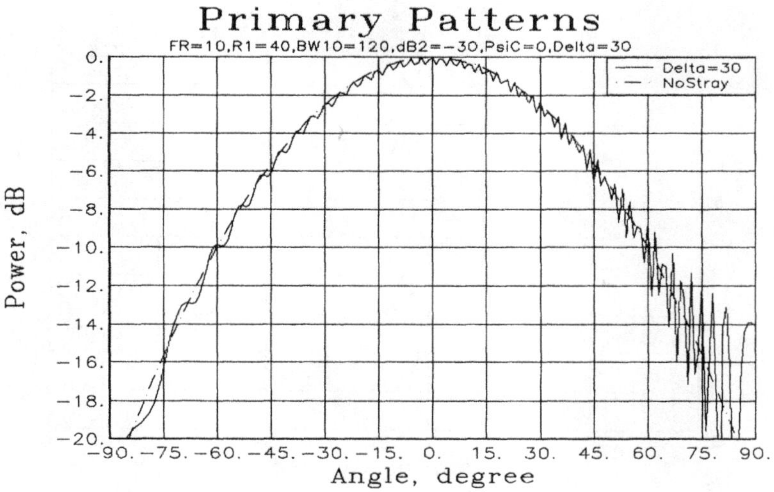

Figure 10.6 Patterns with and without a stray reflecting source located 30 in. from the feed.

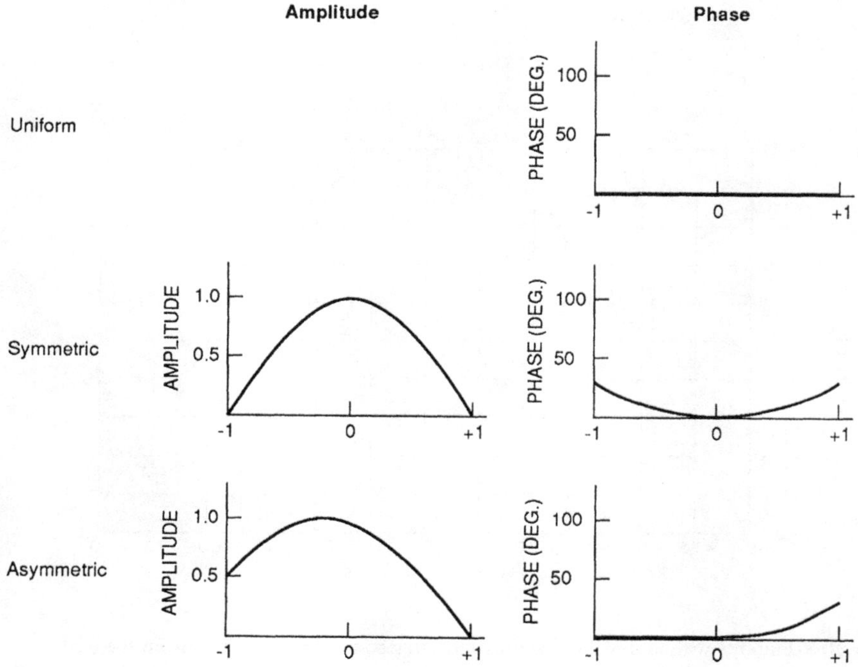

Figure 10.7 Aperture distributions used for calculating far-field patterns.

exists near the feed. The cause of the asymmetry should be corrected (or, at least, understood).

In Figures 10.5 and 10.6, the stray source is 10 and 30 in., respectively, from the phase center of the feed. Note that as δ increases, the angular distance between ripple peaks decreases. Also, the stray source is on the side having the strongest and most "rapid" ripples.

10.3 SECONDARY PATTERNS

Also of interest is how amplitude and phase distributions across the aperture affect the far-field patterns. For example, patterns from line sources have been calculated for various combinations of amplitude (symmetric and asymmetric) and phase (uniform, symmetric, and asymmetric) distributions. The aperture distributions are shown in Figure 10.7, and the far-field patterns are shown in Figure 10.8. The pat-

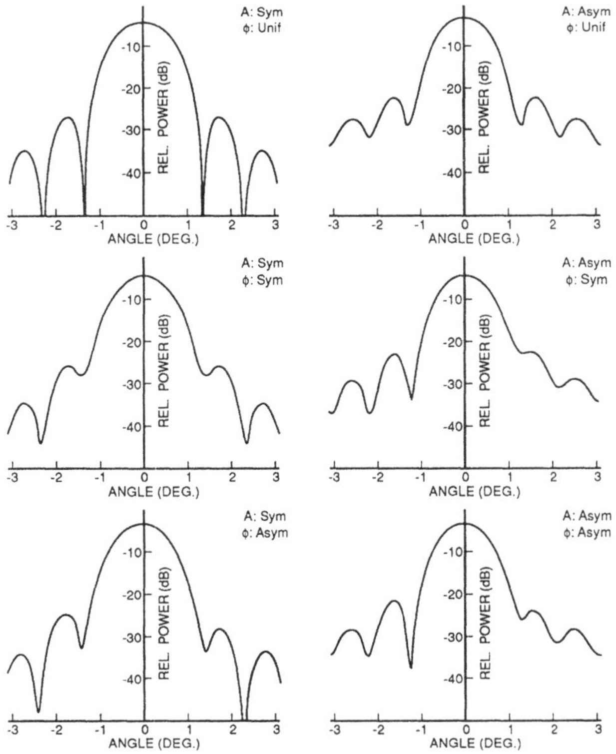

Figure 10.8 Calculated far-field patterns for the indicated aperture distributions.

terns were not normalized for peak power equal to 0 dB because we were interested only in the pattern symmetry and null characteristics, which are summarized in Table 10.1.

Table 10.1
Far-Field Pattern Characteristics (Symmetry and Nulls) for the Indicated Types of Amplitude and Phase Aperture Distributions

Amplitude / *Phase*	*Symmetric*	*Asymmetric*
Uniform	Symmetric Nulls	Symmetric Null-fill
Symmetric	Symmetric Null-fill	Asymmetric Null-fill
Asymmetric	Asymmetric Null-fill	Asymmetric Null-fill

Notice that the only patterns with deep nulls have symmetric amplitude and uniform phase distributions; this is why symmetric antennas are usually focused for deepest nulls on each side of the main lobe. Another interesting observation is that asymmetric amplitude and uniform phase distributions produce a symmetric pattern.

LIST OF SYMBOLS

BW_{10} 10-dB beamwidth of primary pattern
DUT device under test
dB_θ power level at pattern angle θ, dB
dB_2 power level of the radiation from the stray reflecting source relative to the radiation from the feed on axis, dB
dB_t resultant power level, dB
E_1, E_2 relative field strength at probe due to feed and due to stray reflecting source
E_t resultant relative field strength
FR, f frequency
f_1, f_2 the frequencies at two adjacent VSWR peaks
L distance between two reflecting discontinuities
n an integer
R_1, R_2 distances to probe from feed and from stray source
δ distance from feed to stray reflecting source
θ pattern angle

λ	wavelength
λ_{g1}, λ_{g2}	guide wavelengths corresponding to f_1 and f_2
ψ	phase of E_2 relative to the phase of E_1
ψ_c	arbitrary phase-shift constant

APPENDIX

A.1 FREQUENCY BANDS

Table A.1
Designation of Frequency Bands

Radar Bands[a]		*Electronic-Countermeasures Bands*[b]	
Band	*Frequencies*	*Band*	*Frequencies*
HF	3–30 MHz	A	0–250 MHz
VHF	30–300 MHz	B	250–500 MHz
UHF	300–1000 MHz	C	500–1000 MHz
L	1–2 GHz	D	1–2 GHz
S	2–4 GHz	E	2–3 GHz
C	4–8 GHz	F	3–4 GHz
X	8–12 GHz	G	4–6 GHz
Ku	12–18 GHz	H	6–8 GHz
K	18–27 GHz	I	8–10 GHz
Ka	27–40 GHz	J	10–20 GHz
V	40–75 GHz	K	20–40 GHz
W	75–110 GHz	L	40–60 GHz
mm	110–300 GHz	M	60–100 GHz

[a]From IEEE Standard 521-1984.
[b]From AFR 55-44, AR 105-86, OPNAVINST 3430.9B, MCO 3430.1, October 27, 1964.

A.2 PROPERTIES OF DIELECTRICS

Table A.2
Properties of Dielectrics*

		Frequency f, Hz; values of tan δ multiplied by 10^4				
		$f = 10^2$	$f = 10^4$	$f = 10^6$	$f = 10^8$	$f = 10^{10}$
Ceramics						
Alsimag 243	ϵ'/ϵ_0	6.30	6.28	6.22	6.10	5.76
	tan δ	12.5	4.0	3.7	3	8.5
Alsimag 393	ϵ'/ϵ_0	4.95	4.95	4.95	4.95	4.95
	tan δ	38	16	10	10	9.7
Aluminum oxide; Coors AI-200	ϵ'/ϵ_0	8.83	8.82	8.80	8.80	8.79
	tan δ	14	4.8	3.3	3.0	18
Beryllium oxide	ϵ'/ϵ_0	4.61	4.41	4.28	....	4.20
	tan δ	170	74	38	....	5
Magnesium oxide	ϵ'/ϵ_0	9.65	9.65	9.65	9.65	....
	tan δ	3	3	3	3	....
Porcelain, dry-process	ϵ'/ϵ_0	5.50	5.23	5.08	5.02	4.74
	tan δ	220	105	75	98	156
Silicon nitride	ϵ'/ϵ_0	....	....	....	....	5.50
	tan δ	....	....	....	....	30
Steatite 410	ϵ'/ϵ_0	5.77	5.77	5.77	5.77	5.7
	tan δ	55	16	7	6	22
Titania	ϵ'/ϵ_0	100	100	100	100	90
	tan δ	23	6.2	3	2.5	20
Glasses						
Fused quartz	ϵ'/ϵ_0	3.78	3.78	3.78	3.78	3.78
	tan δ	8.5	6	2	1	1
Fused silica	ϵ'/ϵ_0	3.78	3.78	3.78	3.78	3.78
	tan δ	6.6	1.1	0.1	0.3	1.7
E glass	ϵ'/ϵ_0	6.43	6.39	6.32	6.22	6.11
	tan δ	42	27	15	23	60
Low-density materials						
Eccofoam S	ϵ'/ϵ_0	1.05	....	1.18	....	1.47
	tan δ	20	....	40	....	70
Hexcel HRP, 1/4–4.5; honeycomb	ϵ'/ϵ_0	....	....	....	....	1.09
	tan δ	....	....	....	....	30
Hexcel HRH-10, 3/16–4.0; honeycomb	ϵ'/ϵ_0	....	....	....	....	1.12
	tan δ	....	....	....	....	30
Styrofoam 103.7	ϵ'/ϵ_0	1.03	1.03	1.03	....	1.03
	tan δ	2	1	2	....	1.5
Plastics						
Bakelite	ϵ'/ϵ_0	8.2	6.5	5.4	4.4	3.52
	tan δ	1350	630	600	770	366
Duroid 5650	ϵ'/ϵ_0	....	....	....	....	2.65
	tan δ	....	....	....	....	30

Table A.2
Properties of Dielectrics* continued

		Frequency *f*, Hz; values of tan δ multiplied by 10^4				
		$f = 10^2$	$f = 10^4$	$f = 10^6$	$f = 10^8$	$f = 10^{10}$
Epoxy resin RN-48	ϵ'/ϵ_0	3.64	3.61	3.52	3.32	2.91
	tan δ	31	68	142	264	184
Fiberglass; laminated BK-174	ϵ'/ϵ_0	14.2	7.2	5.3	4.8	4.37
	tan δ	2500	1600	460	260	360
Lexan	ϵ'/ϵ_0	....	....	....	....	2.86
	tan δ	....	....	....	....	60
Plexiglas	ϵ'/ϵ_0	3.40	2.95	2.76	....	2.59
	tan δ	605	300	140	....	67
Polystyrene	ϵ'/ϵ_0	2.56	2.56	2.56	2.55	2.54
	tan δ	0.5	0.5	0.7	1	4.3
Rexolite 1422	ϵ'/ϵ_0	2.55	2.55	2.55	2.55	2.54
	tan δ	2.1	1	1.3	3.8	4.7
Polyethylene	ϵ'/ϵ_0	2.25	2.25	2.25	2.25	2.24
	tan δ	2	2	2	2	6.6
Polyvinyl chloride, W-176	ϵ'/ϵ_0	6.21	4.70	3.53	3.00	....
	tan δ	730	1070	720	500	....
Teflon	ϵ'/ϵ_0	2.1	2.1	2.1	2.1	2.08
	tan δ	5	3	2	2	3.7

*From Bodnar [1].
NOTE: Properties are typical for a class of materials. Consult manufacturers for exact properties of their material.

A.3 CONDUCTIVITY OF VARIOUS METALS

Table A.3
Conductivity of Various Metals*

Material	*Conductivity, s/m at 20°C*
Aluminum, commercial hard-drawn	3.54×10^7
Brass, yellow	1.56×10^7
Copper, annealed	5.80×10^7
Copper, beryllium	1.72×10^7
Gold, pure drawn	4.10×10^7
Iron, 99.98 percent pure	1.0×10^7
Iron, gray cast	$0.05 - 0.20 \times 10^7$
Steel	$0.5 - 1.0 \times 10^7$
Lead	0.48×10^7
Nickel	1.28×10^7
Silver, 99.98 percent pure	6.14×10^7
Tin	0.869×10^7
Tungsten, cold-worked	1.81×10^7
Zinc	1.74×10^7
Titanium	0.182×10^7

*From Bodnar [1].

A.4 MULTIPLE AND SUBMULTIPLE PREFIXES

Table A.4
Multiple and Submultiple Prefixes

Multiple/Submultiple	*Prefix*	*Symbol*	*Pronunciation*
10^{12}	tera	T	tĕr′ ȧ
10^{9}	giga	G	jĭ′ gȧ
10^{6}	mega	M	mĕg′ ȧ
10^{3}	kilo	k	kĭl′ o
10^{2}	hecto	h	hĕk′ tō
10	deka	da	dĕk′ ȧ
10^{-1}	dẹci	d	dĕs′ ĭ
10^{-2}	centi	c	sĕn′ tĭ
10^{-3}	milli	m	mĭl′ ĭ
10^{-6}	micro	μ	mī′ krō
10^{-9}	nano	n	năn′ o
10^{-12}	pico	p	pē′ cō
10^{-15}	femto	f	fĕm′ tō
10^{-18}	atto	a	ăt′ tō

A.5 FUNDAMENTAL PHYSICAL CONSTANTS

ϵ_0 permittivity of free space = 8.854×10^{-12} F/m
μ_0 permeability of free space = $4\pi \times 10^{-7}$ H/m
c velocity of light (free space) = 2.997925×10^{8} m/s = 11.803×10^{9} in./s
e electron charge = 1.6021×10^{-19} C
h Planck's constant = 6.6256×10^{-34} J·s

A.6 GREEK ALPHABET

Table A.5
Greek Alphabet*

Name	*Upper-Case*	*Lower-Case*	*Commonly Used to Designate*
Alpha	Α	α	Angles, coefficients, attenuation constant, absorption factor, area
Beta	Β	β	Angles, coefficients, phase constant
Gamma	Γ	γ	Complex propagation constant (cap), specific gravity, angles, electrical conductivity, propagation constant
Delta	Δ	δ	Increment or decrement (cap or small), determinant (cap), permittivity (cap), density, angles
Epsilon	Ε	ϵ	Dielectric constant, permittivity, base of natural logarithms, electric intensity
Zeta	Ζ	ζ	Coordinates, coefficients
Eta	Η	η	Intrinsic impedance, efficiency, surface charge density, hysteresis, coordinates
Theta	Θ	ϑ, θ	Angular phase displacement, time constant, reluctance, angles
Iota	Ι	ι	Unit vector
Kappa	Κ	κ	Susceptibility, coupling coefficient, thermal conductivity
Lambda	Λ	λ	Permeance (cap), wavelength, attenuation constant
Mu	Μ	μ	Permeability, amplification factor, prefix micro
Nu	Ν	ν	Reluctivity, frequency
Xi	Ξ	ξ	Coordinates
Omicron	Ο	o	
Pi	Π	π	3.1416
Rho	Ρ	ρ	Resistivity, volume charge density, coordinates
Sigma	Σ	σ	Summation (cap), surface charge density, complex propagation constant, electrical conductivity, leakage coefficient, deviation
Tau	Τ	τ	Time constant, volume resistivity, time-phase displacement, transmission factor, density
Upsilon	Υ	υ	
Phi	Φ	ϕ, φ	Scalar potential (cap), magnetic flux, angles
Chi	Χ	χ	Electric susceptibility, angles
Psi	Ψ	ψ	Dielectric flux, phase difference, coordinates, angles
Omega	Ω	ω	Resistance in ohms (cap), solid angle (cap), angular velocity

*From Mechtly [2].

A.7 DECIBELS AND NEPERS

The decibel (dB) is a dimensionless unit for representing the ratio of two values of power. The number n of decibels is 10 times the logarithm to the base 10 of the power ratio:

$$n(\text{dB}) = 10 \log(P_2/P_1) \tag{A.1}$$

The neper (Np) is a dimensionless unit for representing the ratio of two values of amplitude. The number n of nepers is the natural logarithm of the amplitude ratio:

$$n(\text{Np}) = \ln(A_2/A_1) = \frac{1}{2} \ln(P_2/P_1) \tag{A.2}$$

To convert decibels to nepers, multiply by 0.1151; to convert nepers to decibels, multiply by 8.686.

A.8 RECTANGULAR WAVEGUIDES AND FLANGES

Table A.6
Standard Rectangular Waveguides and Flanges*

EIA Designation WR-	DOD Part No. M85/1-	AN Designation	Material	Inside Dimensions, in. (mm)	Tolerance, in. (mm)	Wall Thickness, in. (mm)	Frequency Range, GHz for TE_{10}	f_c GHz for TE_{10}	λ_c mm	Lowest and highest: Attenuation dB/100 ft (dB/30.5m)	Lowest and highest: Theoretical Peak Power, MW	Lowest and highest: Theoretical Maximum Continuous Wave, kW	Flanges: Cover	Flanges: Choke
650	017-	69/U	B	6.500 × 3.250	±0.008	0.080	1.12–1.70	0.908	330.2	0.316–.0209	41.3–59.7	80.5–122	323/U	322/U
	018-	103/U	A	(165.1 × 82.55)	(0.20)	(2.03)				0.273–0.180		88.5–136	1720/U[a]	
510	023-	337/U	B	5.100 × 2.550	±0.008	0.080	1.45–2.20	1.154	259.1	0.440–0.299	26.2–37.0	47.9–70.4	1715/U[a]	
	025-	338/U	A	(129.5 × 64.77)	(0.20)	(2.03)				0.380–0.258		53.2–78.3	1717/U[a]	
430	029-	105/U	A	4.300 × 2.150	±0.008	0.080	1.70–2.60	1.375	218.4	0.502–0.334	18.2–26.3	35.3–53.1	437/U[a]	
	031-	104/U	B	(109.2 × 54.61)	(0.20)	(2.03)				0.583–0.387		31.7–47.7	435A/U[a]	
340	035-	113/U	A	3.400 × 1.700	±0.006	0.080	2.20–3.30	1.737	172.7	0.682–0.474	11.9–16.4	21.7–31.3	554/U[a]	
	037-	112/U	B	(86.36 × 43.18)	(0.15)	(2.03)				0.791–0.550		19.5–28.1	553A/U[a]	
284	041-	75/U	A	2.840 × 1.340	±0.006	0.080	2.60–3.95	2.080	144.3	0.950–0.651	7.65–10.9	13.4–19.6	584/U	585A/U
	043-	48/U	B	(72.14 × 34.04)	(0.15)	(2.03)				1.10–0.754		12.1–17.6	53/U	54B/U
229	047-	341/U	A	2.290 × 1.145	±0.006	0.064	3.30–4.90	2.577	116.3	1.21–0.858	5.48–7.55	8.99–12.7	1727/U[a]	
	049-	340/U	B	(58.16 × 29.08)	(0.15)	(1.63)				1.40–0.996		8.08–11.4	1726/U[a]	
187	053-	95/U	A	1.872 × 0.872	±0.005	0.064	3.95–5.85	3.155	95.10	1.79–1.24	3.30–4.70	5.17–7.45	407/U	406B/U
	055-	49/U	B	(47.55 × 22.15)	(0.13)	(1.63)				2.07–1.44		4.64–6.69	149A/U	148C/U
159	059-	344/U	A	1.590 × 0.795	±0.005	0.064	4.90–7.05	3.705	80.77	1.99–1.49	2.79–3.72	4.20–5.62	1731/U[a]	
	061-	343/U	B	(40.39 × 20.19)	(0.13)	(1.63)				2.31–1.72		3.77–5.05	247/U	248/U
137	065-	106/U	A	1.372 × 0.622	±0.004	0.064	5.85–8.20	4.285	69.70	2.53–2.00	1.98–2.53	2.90–3.67	441/U	440B/U
	067-	50/U	B	(34.85 × 15.80)	(0.10)	(1.63)				2.94–2.32		2.60–3.29	344B/U	343B/U
112	071-	68/U	A	1.122 × 0.497	±0.004	0.064	7.05–10.0	5.260	57.00	3.55–2.76	1.28–1.70	1.79–2.30	138/U	137B/U
	073-	51/U	B	(28.50 × 12.62)	(0.10)	(1.63)				4.11–3.20		1.61–2.07	51/U	52B/U
90	077-	67/U	A	0.900 × 0.400	±0.004	0.050	8.20–12.4	6.560	45.72	5.54–3.83	0.758–1.12	0.959–1.39	135/U	136B/U
	079-	52/U	B	(22.86 × 10.16)	(0.10)	(1.27)				6.42–4.45		0.862–1.25	39/U	40B/U
75	083-	347/U	A	0.750 × 0.375	±0.003	0.050	10.0–15.0	7.847	38.10	6.55–4.58	0.622–0.903	0.737–1.06		
	085-	346/U	B	(19.05 × 9.53)	(0.08)	(1.27)				7.60–5.31		0.662–0.948		
62	089-	91/U	B	0.622 × 0.311	±0.0025	0.040	12.4–18.0	9.490	31.60	9.58–7.04	0.457–0.633	0.451–0.614	419/U	541A/U
	090-	349/U	A	(15.80 × 7.90)	(0.06)	(1.02)				8.26–6.07		0.502–0.683	1665/U	1666/U
	093-	107/U	SC							6.91–5.08		0.602–0.818	419/U	541/U
51	096-	353/U	B	0.510 × 0.255	±0.0025	0.040	15.0–22.0	11.54	25.91	13.1–9.48	0.312–0.433	0.290–0.400		
	097-	351/U	A	(12.95 × 6.48)	(0.06)	(1.02)				11.3–8.17		0.323–0.445		
42	102-	53/U	B	0.420 × 0.170	±0.002	0.040	18.0–26.5	14.08	21.34	20.5–15.0	0.171–0.246	0.157–0.213	595/U	596A/U
	103-	121/U	A	(10.67 × 4.32)	(0.05)	(1.02)				17.7–13.0		0.174–0.237	597/U	598A/U
	106-	66/U	SC							14.8–10.9		0.209–0.284	595/U	596A/U

34	109-	354/U	B	0.340 × 0.170	±0.002	0.040	22.0–33.0	17.28	17.27	25.0–17.4	0.139–0.209	0.118–0.169	1530/U[a]	
	110-	355/U	A	(8.64 × 4.32)	(0.05)	(1.02)				21.6–15.0		0.131–0.188		
	113-	357/U	SC							16.2–11.3		0.175–0.252		
28	114-	96/U	S	0.280 × 0.140	±0.0015	0.040	26.5–40.0	21.10	14.22	24.6–16.8	0.096–0.146	0.212–0.310	599/U	600A/U
	117-	271/U	SC	(7.11 × 3.56)	(0.04)	(1.02)				22.0–15.1		0.242–0.347		
22	118-	97/U	S	0.224 × 0.112	±0.0010	0.040	33.0–50.0	26.35	11.38	34.5–23.5	0.064–0.097	0.134–0.197	383/U[a]	
	121-	272/U	SC	(5.69 × 2.84)	(0.03)	(1.02)				31.0–21.1		0.149–0.219		
19	124-	358/U	SC	0.188 × 0.094	±0.0010	0.040	40.0–60.0	30.69	9.550	38.0–27.3	0.048–0.070	0.111–0.155	1529/U[a]	
				(4.78 × 2.39)	(0.03)	(1.02)								
15	125-	98/U	S	0.148 × 0.074	±0.0010	0.040	50.0–75.0	39.90	7.518	64.2–43.9	0.030–0.044	0.052–0.077	385/U[a]	
	128-	273/U	SC	0(3.76 × 1.88)	(0.03)	(1.02)				57.6–39.3		0.058–0.085		
12	129-	99/U	S	0.122 × 0.061	±0.0005	0.040	60.0–90.0	48.40	6.198	87.8–58.9	0.020–0.030	0.047–0.067	387/U[a]	
	132-	274/U	SC	(3.10 × 1.55)	(0.01)	(1.02)				78.7–52.7		0.049–0.074		
10	135-	359/U	SC	0.100 × 0.050	±0.0005	0.040	75.0–110	58.85	5.080	101–71.0	0.014–0.020	0.032–0.045	1528/U[a]	
				(2.54 × 1.27)	(0.01)	(1.02)								
8	138-	278/U	SC	0.080 × 0.040	±0.0003	0.020	90.0–140	73.84	4.064	154–98.7	0.009–0.013	0.015–0.024	1527/U[a]	
				(2.03 × 1.02)	(0.01)	(0.51)								
7	141-	276/U	SC	0.065 × 0.0325	±0.00025	0.020	110–170	90.85	3.302	214–135	0.006–0.009	0.010–0.016	1525/U[a]	
				(1.65 × 0.83)	(0.01)	(0.51)								
5	144-	275/U	SC	0.051 × 0.0255	±0.00025	0.020	140–220	115.8	2.591	308–194	0.004–0.006	0.006–0.01	1524/U[a]	
				(1.30 × 0.65)	(0.01)	(0.51)								
4	147-	277/U	SC	0.043 × 0.0215	±0.0002	0.020	170–260	137.5	2.184	377–251	0.003–0.005	0.005–0.007	1526/U[a]	
				(1.09 × 0.546)	(0.01)	(0.51)								
3	152-	139/U	S	0.034 × 0.0170	±0.0002	Round	220–325	173.3	1.727	512–341	0.0004–0.0005	0.005–0.008		
				(0.864 × 0.432)	(0.01)									

*From Lowman [3].

Materials	*Resistivity, μΩ · cm*
A = aluminum alloy 1100	2.90
B = brass	3.90
SC = silver-clad copper	
S = silver	1.63

[a]These flanges mate with themselves.

A.9 VSWR AND RELATED TERMS

Table A.7
The Reflection Coefficient, Return Loss, and Transmission Loss as a Function of VSWR

VSWR	*Reflection Coefficient*	*Return Loss, dB*	*Transmission Loss, dB*	*VSWR*	*Reflection Coefficient*	*Return Loss, dB*	*Transmission Loss, dB*
1.01	0.0050	46.06	0.00	3.10	0.5122	5.81	1.32
1.02	0.0099	40.09	0.00	3.20	0.5238	5.62	1.39
1.03	0.0148	36.61	0.00	3.30	0.5349	5.43	1.46
1.04	0.0196	34.15	0.00	3.40	0.5455	5.26	1.53
1.05	0.0244	32.26	0.00	3.50	0.5556	5.11	1.60
1.06	0.0291	30.71	0.00	3.60	0.5652	4.96	1.67
1.07	0.0338	29.42	0.00	3.70	0.5745	4.81	1.74
1.08	0.0385	28.30	0.01	3.80	0.5833	4.68	1.81
1.09	0.0431	27.32	0.01	3.90	0.5918	4.56	1.87
1.10	0.0476	26.44	0.01	4.00	0.6000	4.44	1.94
1.12	0.0566	24.94	0.01	4.20	0.6154	4.22	2.07
1.14	0.0654	23.69	0.02	4.40	0.6296	4.02	2.19
1.16	0.0741	22.61	0.02	4.60	0.6429	3.84	2.32
1.18	0.0826	21.66	0.03	4.80	0.6552	3.67	2.44
1.20	0.0909	20.83	0.04	5.00	0.6667	3.52	2.55
1.25	0.1111	19.08	0.05	5.20	0.6774	3.38	2.67
1.30	0.1304	17.69	0.07	5.40	0.6875	3.25	2.78
1.35	0.1489	16.54	0.10	5.60	0.6970	3.14	2.89
1.40	0.1667	15.56	0.12	5.80	0.7059	3.03	3.00

1.45	0.1837	14.72	0.15	6.00	0.7143	2.92	3.10
1.50	0.2000	13.98	0.18	6.20	0.7222	2.83	3.20
1.55	0.2157	13.32	0.21	6.40	0.7297	2.74	3.30
1.60	0.2308	12.74	0.24	6.60	0.7368	2.65	3.40
1.65	0.2453	12.21	0.27	6.80	0.7436	2.57	3.50
1.70	0.2593	11.73	0.30	7.00	0.7500	2.50	3.59
1.75	0.2727	11.29	0.34	7.50	0.7647	2.33	3.82
1.80	0.2857	10.88	0.37	8.00	0.7778	2.18	4.03
1.85	0.2982	10.51	0.40	8.50	0.7895	2.05	4.24
1.90	0.3103	10.16	0.44	9.00	0.8000	1.94	4.44
1.95	0.3220	9.84	0.48	9.50	0.8095	1.84	4.63
2.00	0.3333	9.54	0.51	10.00	0.8182	1.74	4.81
2.10	0.3548	9.00	0.58	11.00	0.8333	1.58	5.15
2.20	0.3750	8.52	0.66	12.00	0.8462	1.45	5.47
2.30	0.3939	8.09	0.73	13.00	0.8571	1.34	5.76
2.40	0.4118	7.71	0.81	14.00	0.8667	1.24	6.04
2.50	0.4286	7.36	0.88	15.00	0.8750	1.16	6.30
2.60	0.4444	7.04	0.96	20.00	0.9048	0.87	7.41
2.70	0.4595	6.76	1.03	25.00	0.9231	0.70	8.30
2.80	0.4737	6.49	1.10	30.00	0.9355	0.58	9.04
2.90	0.4872	6.25	1.18	35.00	0.9444	0.50	9.66
3.00	0.5000	6.02	1.25	40.00	0.9512	0.43	10.21

Note: Losses given in decibels.

A.10 EXPANDED-METAL MESH REFLECTORS

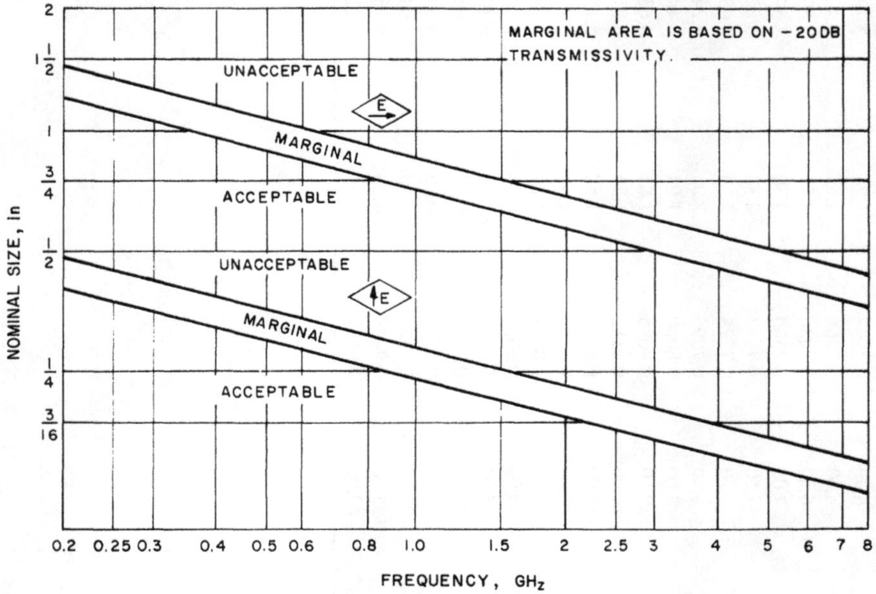

Figure A.1 Transmission (leakage) through expanded-metal mesh reflectors. (From [4].)

A.11 WIND LOADING ON A PARABOLOIDAL REFLECTOR

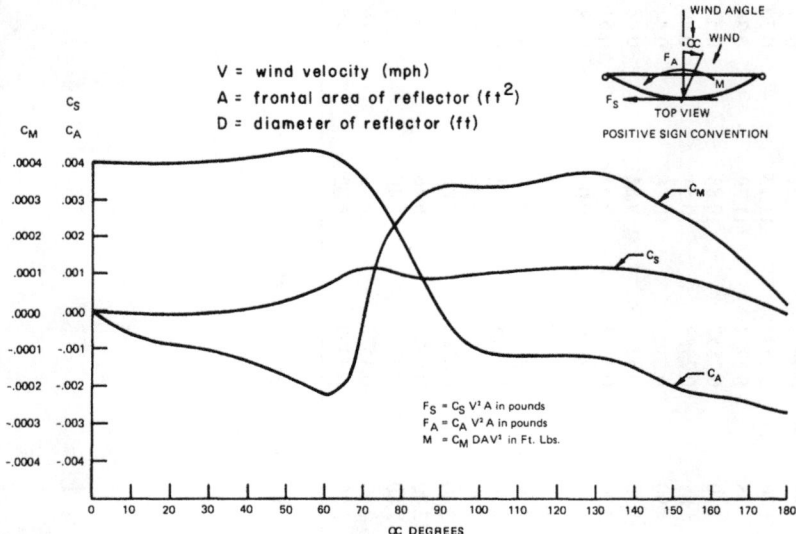

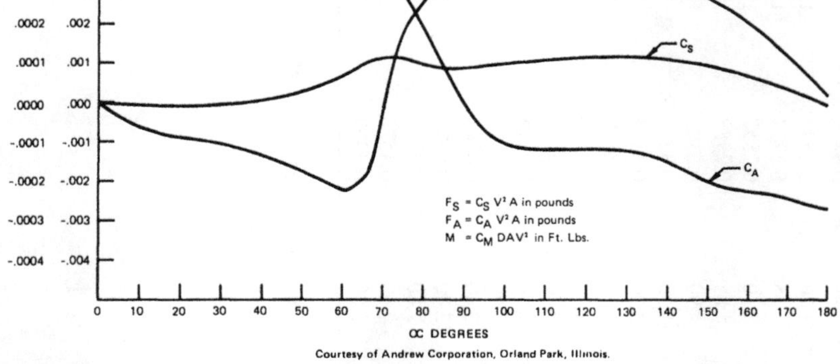

Figure A.2 Wind forces and torques on a paraboloidal reflector. (From Saad [5].)

A.12 EFFECTS OF APERTURE BLOCKAGE

The positioning of a feed, support struts, or transmission lines in front of an aperture antenna generally will reduce the gain and raise the maximum sidelobes. In a simple but useful geometric-optics approximation, the blockage is assumed to produce a "hole" in the aperture distribution. The case of a circular reflector aperture of diameter D and a circular feed blockage of diameter D_b is illustrated by Hannan [6] in Figure A.3. The resulting illumination is the sum of two components [7]: the original (unblocked) illumination and the negative hole (blocked) illumination. The resulting pattern is the sum of the original pattern plus the hole pattern.

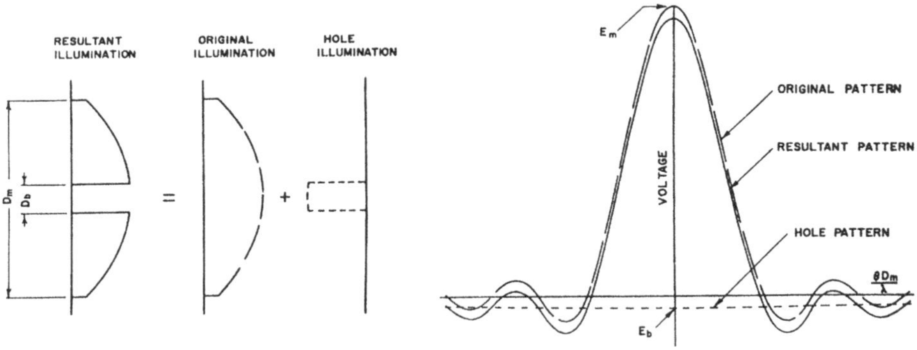

Figure A.3 Effects of aperture blockage. (©1961 IEEE, from Hannan [6].)

Bodnar [1] reports that using a line source of length L with a $\cos^n$ aperture distribution and a central blockage of L_b produces the gain losses and sidelobe degradations shown in Figure A.4. The use of a circular aperture of diameter D and a parabola to a power aperture distribution blocked by a central disk of diameter D_b produces the gain losses and sidelobe degradations shown in Figure A.5.

The graphical data of Figures A.4 and A.5 allow us to estimate quickly gain losses and sidelobe degradations for central blockages; a more detailed analysis for reflector antennas having both central feed blockage and support-strut blockage was reported by Gray [8]. In the far field, the maximum field intensity of the hole pattern relative to the maximum field intensity of the unperturbed aperture is

$$\frac{E_h}{E} \approx 2\left(\frac{D_b}{D}\right)^{1/2} \tag{A.3}$$

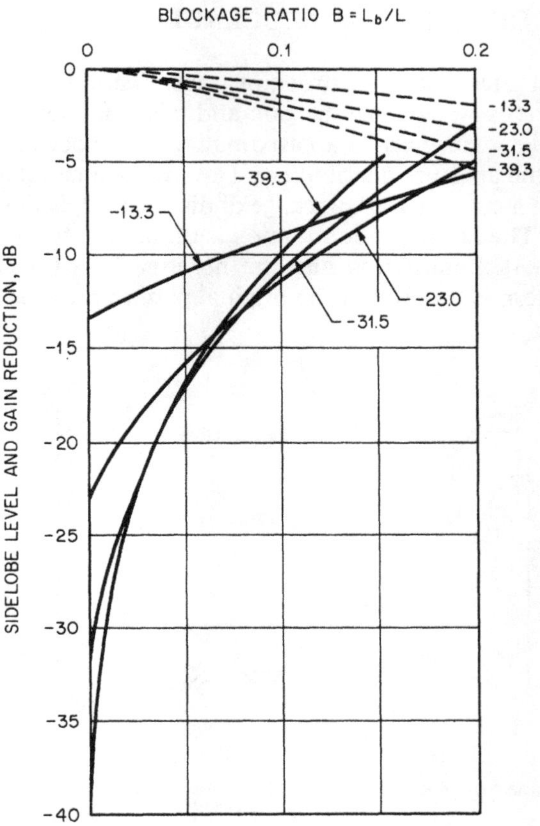

Figure A.4 Gain reduction (----) and resulting sidelobe level (———) for a centrally blocked line-source distribution having the indicated unblocked sidelobe level. (From Bodnar [1].)

Consider a pair of struts of width W that block a narrow band across the reflector's diameter; the maximum field intensity of the strut pattern relative to the maximum field intensity of the unperturbed aperture is

$$\frac{E_s}{E} \approx 1.55 \frac{W}{D} \tag{A.4}$$

A strut pattern will be fan-shaped with the narrow beamwidth approximately equal to the reflector beamwidth. Thus, when we employ more than one pair of struts, the strut patterns will add together to reduce antenna gain, but they will not add together in the sidelobe regions.

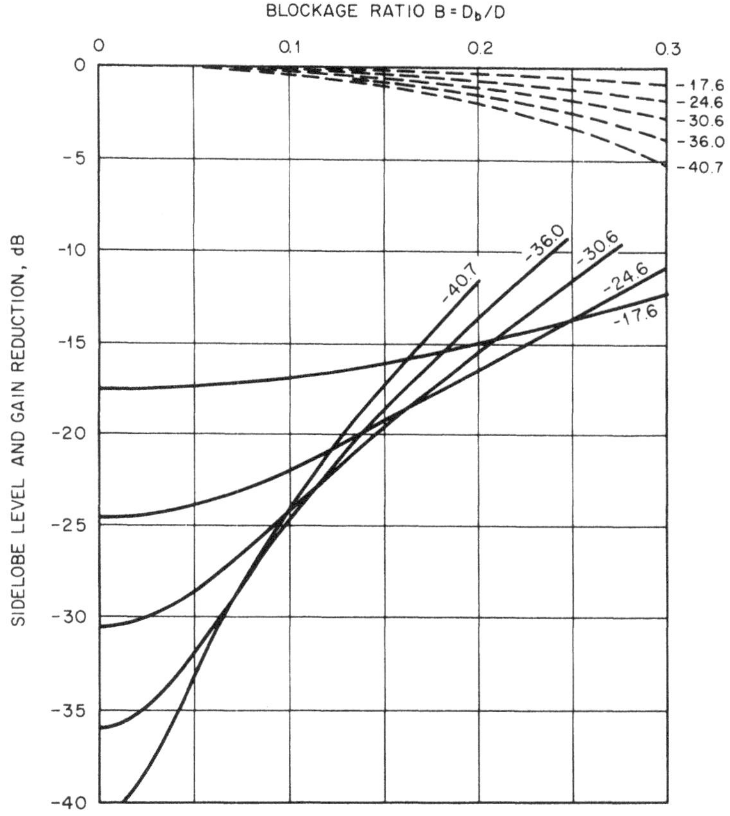

Figure A.5 Gain reduction (----) and resulting sidelobe level (———) for a centrally blocked circular-aperture distribution having the indicated unblocked sidelobe level. (From Bodnar [1].)

Gray works through an example of a 30-foot-diam dish reflector with three pairs of 3-in.-diam struts. The relative field intensity of a single pair of struts is 0.0129; on axis, the three pairs of struts add to produce a relative field intensity of 0.0387. The central feed diameter is 23.5 in.; thus, the relative field intensity of the hole pattern is 0.0085. The unperturbed aperture has a theoretical sidelobe level of −30.6 dB, which equates to a relative field intensity of 0.0295.

The perturbed pattern has a main-lobe field intensity of 1.0000 − 0.0387 − 0.0085 = 0.9528, which is a gain reduction of 0.4 dB. The perturbed sidelobe level is 0.0295 + 0.0129 + 0.0085 = 0.0509; referenced to the perturbed main lobe, this yields a perturbed sidelobe level of −25.4 dB. The reflector manufacturing errors are so small (in Gray's example) relative to a wavelength that they have negligible effects on the sidelobe levels or main-lobe gain. The measured maximum sidelobes are -24.5 to -30 dB, which are quite close to the predicted sidelobes.

A.13 MATHEMATICAL MODELS FOR MAIN LOBES

Mathematical models of the main lobes of radiation patterns are often convenient and desirable. This section will consider some of the popular models; for comparison, the various beams will be made coincident at a specific relative power level dB_c and pattern angle θ_c.

A.13.1 Parabolic Beam

The parabolic beam was discussed briefly in Chapter 2; the relative power expressed in dB *versus* the pattern angle has a parabolic shape, according to Kelleher [9]. From Chapter 2, the beam pattern in dB can be written as

$$\mathrm{dB(parabola)} = \mathrm{dB}_c\left(\frac{\theta}{\theta_c}\right)^2 = \left[\frac{\mathrm{dB}_c}{\theta_c^2}\right]\theta^2 \tag{A.5}$$

A.13.2 cos-*n*th Beam

The cos-*n*th beam has a field pattern of $\cos^n\theta$ and a power pattern of $\cos^{2n}\theta$. Thus, in a dB plot,

$$\mathrm{dB}(\cos\text{-}n\mathrm{th}) = 20n\,\log(\cos\theta)$$

At the coincidence point,

$$\mathrm{dB}_c = 20n\,\log(\cos\theta_c)$$

Thus, the beam pattern is

$$\mathrm{dB}(\cos\text{-}n\mathrm{th}) = \frac{\mathrm{dB}_c}{\log(\cos\theta_c)}\log(\cos\theta) \tag{A.6}$$

A.13.3 Gaussian Beam

The Gaussian function, normalized to have an area of unity under the curve, is [10]:

$$f(x) = \frac{1}{\sigma\sqrt{2\pi}} \exp\left[-\frac{1}{2}\left(\frac{x - m}{\sigma}\right)^2 \right]$$

where m is the mean and σ is the standard deviation. We want our field pattern to be unity on axis, so the field pattern is

$$F = e^{-K_g\theta^2}$$

where K_g is a constant that we can adjust to set the pattern width. Thus, in a dB plot

$$\text{dB(Gaussian)} = 20 \log (e^{-K_g\theta^2}) = 20 \frac{\ln(e^{(-K_g\theta^2)})}{\ln 10}$$

or

$$\text{dB(Gaussian)} = \frac{-20K_g}{\ln 10}\theta^2$$

At the coincidence point,

$$\text{dB}_c = \frac{-20K_g}{\ln 10}\theta_c^2$$

Thus, the beam pattern is

$$\text{dB(Gaussian)} = \left[\frac{\text{dB}_c}{\theta_c^2}\right]\theta^2 \tag{A.7}$$

which is the same as eq. (A.5). Thus, the Gaussian field pattern becomes a parabola when plotted as dB *versus* angle.

A.13.4 (sin*x*)/*x* Beam

The (sin$x$)/x field pattern is

$$F = \frac{\sin x}{x}$$

where $x = (\pi L/\lambda)\sin\theta = K_s \sin\theta$; the constant K_s can be adjusted to change the beamwidth. In a dB plot,

$$\mathrm{dB}(\sin x/x) = 20\log\left[\frac{\sin(K_s \sin\theta)}{K_s \sin\theta}\right] \tag{A.8}$$

At the coincident point,

$$\mathrm{dB}_c = 20\log\left[\frac{\sin(K_s \sin\theta_c)}{K_s \sin\theta_c}\right]$$

This equation must be solved for the value of K_s (a trial-and-error technique is fairly fast) to be used in eq. (A.8).

The (sin$x$)/x pattern assumes that one is dealing with a narrow beamwidth; for wide beamwidths, an obliquity factor should be included, as in

$$F = \left[\frac{\sin(K_s \sin\theta)}{K_s \sin\theta}\right]\frac{1+\cos\theta}{2}$$

In a dB plot,

$$\mathrm{dB}[(\sin x)/x] = 20\log\left[\frac{\sin(K_s \sin\theta)(1+\cos\theta)}{2K_s \sin\theta}\right] \tag{A.9}$$

This expression for the (sin$x$)/x beam should be satisfactory out to about 30 degrees. At the coincident point,

$$\mathrm{dB}_c = 20\log\left[\frac{\sin(K_s \sin\theta_c)(1+\cos\theta_c)}{2K_s \sin\theta_c}\right]$$

This must be solved for the value of K_s for use in eq. (A.9).

A.13.5 Primary Pattern Comparisons

The above theoretical patterns were compared with a coincident power level of -10 dB and with coincident angles of 60, 30, and 6 degrees; the results are illustrated in Figures A.6 through A.8. Patterns for Gaussian beams are not shown because they are identical to patterns for the parabolic beams.

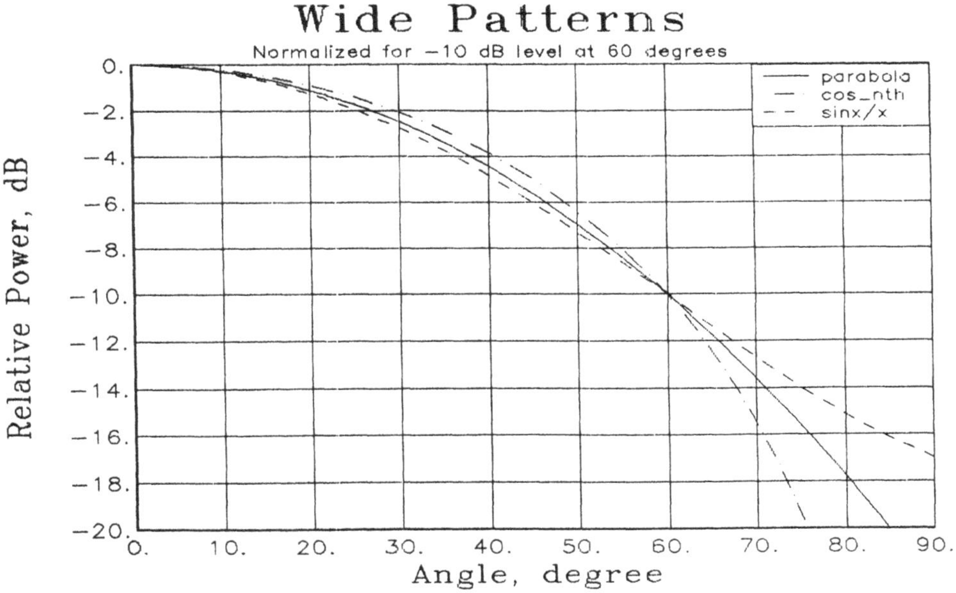

Figure A.6 Comparison of wide-beamwidth pattern models.

The wide beamwidths of Figure A.6 that have 10-dB widths of 120 degrees are typical of primary patterns for front-fed paraboloidal reflectors. The $(\sin x)/x$ pattern is shown, but it is not valid at such wide angles. Note that the cos-nth pattern differs significantly from the parabolic pattern. Kelleher [9] reported that measured feed patterns conform closely to the parabolic pattern except at the lower power levels; the patterns are essentially identical down to -10 dB and then the measured patterns get slightly wider until, at -20 dB, the measured patterns are about 1 dB higher than the parabolic pattern. Wide-beamwidth cos-nth primary patterns probably will predict secondary beamwidths reasonably well, but some

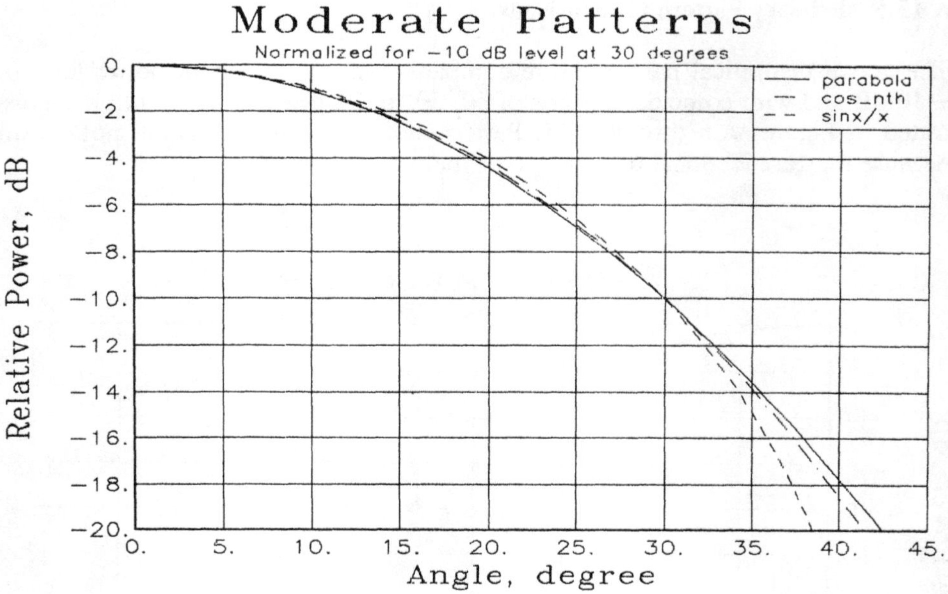

Figure A.7 Comparison of moderate-beamwidth pattern models.

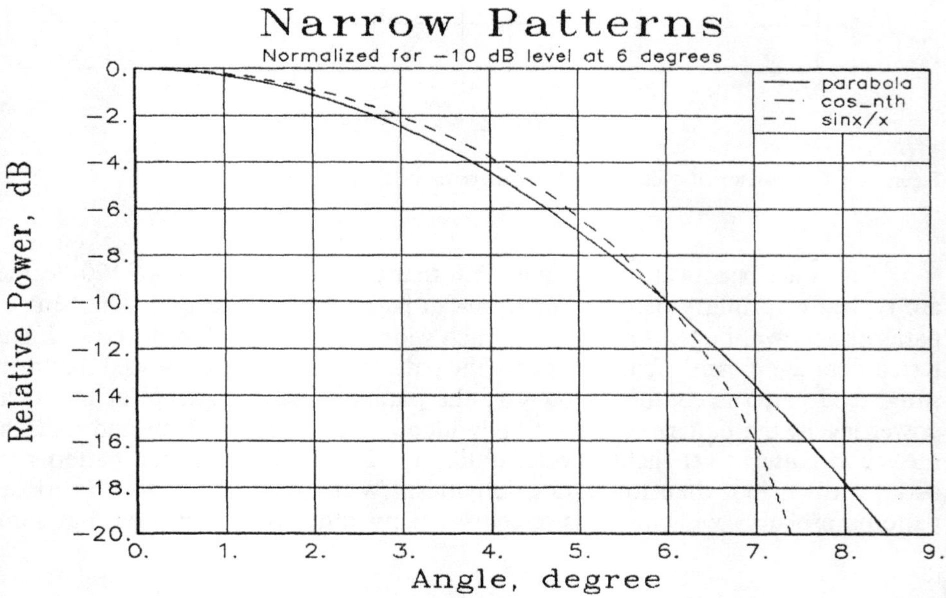

Figure A.8 Comparison of narrow-beamwidth pattern models.

doubt exists about whether they will predict sidelobes satisfactorily—particularly when the edge illumination is low. Based on Kelleher's data [9], parabolic beams seem to be preferred as mathematical models for wide-beamwidth patterns.

The moderate beamwidths of Figure A.7 that have −10-dB widths of 60 degrees are typical of primary patterns for offset-fed paraboloidal reflectors. The $(\sin x)/x$ pattern in Figure A.7 includes the effects of the obliquity factor and, therefore, the pattern should be reasonably accurate. Note that the cos-nth pattern conforms quite closely to the parabolic pattern; thus, the two should predict similar secondary patterns and both are probably acceptable as mathematical models.

The narrow beamwidths of Figure A.8 that have −10-dB widths of 12 degrees are too narrow to simulate primary patterns for most antennas, but they illustrate an interesting behavior. The $(\sin x)/x$ pattern does not include the effects of the obliquity factor because such effects are negligible at these small pattern angles. Note that the cos-nth and parabolic patterns are essentially identical at these beamwidths, but they differ from the $(\sin x)/x$ pattern. Suitable mathematical models for these narrow beams appear to be parabolic (or Gaussian) or cos-nth. The $(\sin x)/x$ models do not appear to be satisfactory for any of the primary patterns that were considered.

The reason for the merging of the cos-nth and parabolic patterns at narrow beamwidths can be seen as follows. The beam pattern was given in eq. (A.6):

$$\mathrm{dB}(\cos\text{ - }n\mathrm{th}) = \frac{\mathrm{dB}_c}{\log(\cos\theta_c)}\log(\cos\theta)$$

This also can be expressed as

$$\mathrm{dB}(\cos\text{ - }n\mathrm{th}) = \frac{\mathrm{dB}_c}{\ln(\cos\theta_c)}\ln(\cos\theta)$$

Use the series expansion,

$$\ln(\cos\theta) = -\frac{\theta^2}{2} - \frac{\theta^4}{12} - \dots$$

At our small angles, only the first term is significant; thus,

$$\mathrm{dB}(\cos\text{ - }n\mathrm{th}) \approx \left[\frac{\mathrm{dB}_c}{\theta_c^2}\right]\theta^2$$

which is the same expression as that for the parabola in eq. (A.5).

A.13.6 Secondary Pattern Comparisons

The above mathematical models were selected as simulations of primary beam patterns. We now compare some mathematical models for narrow far-field patterns that are typical of microwave radar antennas. Figure A.9 illustrates three patterns (parabolic, Taylor, and $(\sin x)/x$) with -10-dB widths of 4 degrees. The parabolic and $(\sin x)/x$ models are the same as those used above for the primary beams, and the Taylor beam was calculated using a 45-element array with -30-dB sidelobes and $\bar{n} = 4$. Note that significant differences exist between any two of the theoretical beams. None of the three mathematical models appears to represent accurately all narrow far-field beams; however, any of the three models probably would be suitable for many applications that do not require great accuracy.

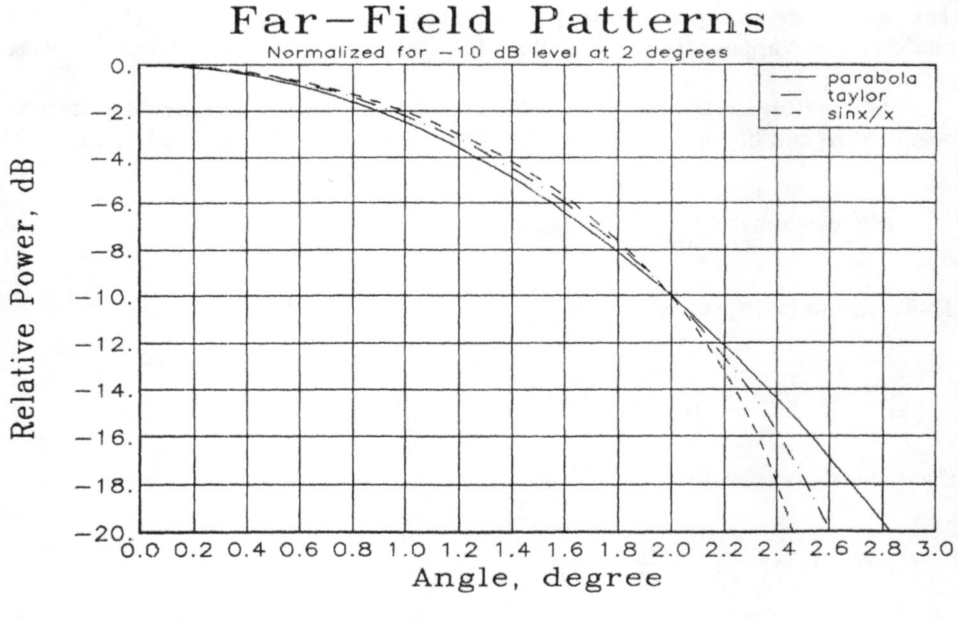

Figure A.9 Comparison of narrow far-field pattern models.

A.14 VSWR—PHASE SHIFT

When a wave in a transmission line encounters a lossless discontinuity, some energy is reflected and the remaining energy is transmitted past the discontinuity.

Most engineers know that such a discontinuity introduces a phase shift in the transmitted wave, but many are not familiar with the rule of thumb that "for small reflections, the phase shift is approximately equal to the reflection coefficient."

I have heard this rule-of-thumb many times, but I have not seen it discussed in the literature. A nonrigorous explanation follows.

Suppose that the incident wave has a field strength of unity. Let

- Γ_r = the field reflection coefficient
- Γ_t = the field transmission coefficient

Conservation of energy requires that $\Gamma_r^2 + \Gamma_t^2 = 1$. Thus, at the discontinuity, the phasors are oriented as indicated in Figure A.10. The phase shift is

$$\Delta\phi = \arcsin\Gamma_r \approx \Gamma_r \tag{A.10}$$

Cotton [11] reported that this relationship was approximately correct in a simple experiment conducted by a student; however, it should be used for rough estimates only.

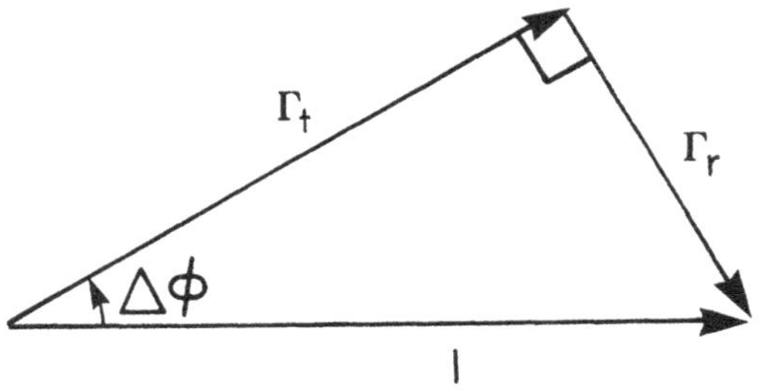

Figure A.10 Phasor orientations.

A.15 RADIATION HAZARDS

As stated by the American National Standards Institute in Reference [12], "Recommendations are made to prevent possible harmful effects in human beings exposed to electromagnetic fields in the frequency range from 300 kHz to 100 GHz. These recommendations are intended to apply to non-occupational as well as to occupational exposures. These recommendations are not intended to apply to the purposeful exposure of patients by or under the direction of practitioners of the healing arts."

The protection guides, in terms of the maximum equivalent plane-wave free-space power density, as a function of frequency are reproduced in Table A.8. A graphic representation of the protection guides is shown in Figure A.11.

Table A.8
Radio-Frequency Protection Guides*

Frequency Range (MHz)	*Power Density (mW/cm²)*
0.3–3	100
3–30	$999/f^2$
30–300	1.0
300–1500	$f/300$
1500–100,000	5.0

*Data are from [12].
Note: f = frequency (MHz)

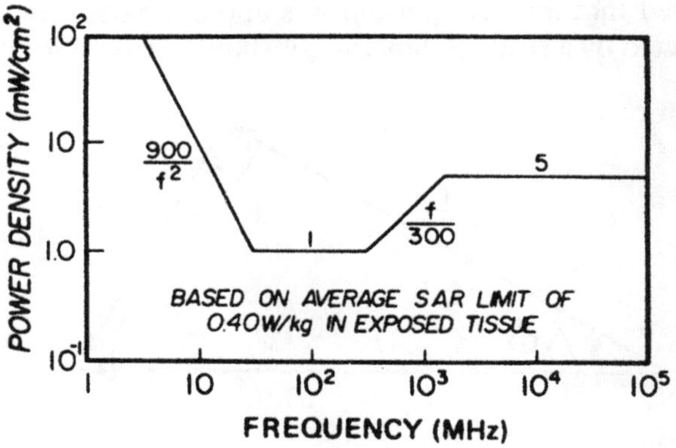

Figure A.11 Radio-frequency protection guide for whole-body exposure of human beings. (Reproduced from [12], ©1982 by the IEEE, with permission of the IEEE Standards Department.)

A.16 SAGITTA—PHASE ERRORS

Figure A.12 depicts a circular arc with radius R, angular width $2\theta_f$, and chord length D. The distance from the center of the circle to the chord is L, and the distance from the center of the chord to the arc is Δ, which is called the *sagitta.*

The sagitta is important when estimating phase errors due to spherical or circular wavefronts such as at the aperture of horns or at the antenna under test on antenna ranges.

Several ways exist to calculate the value of Δ; some of them are

$$\Delta = R - L = R - R\cos\theta_f = R(1 - \cos\theta_f) \tag{A.11}$$

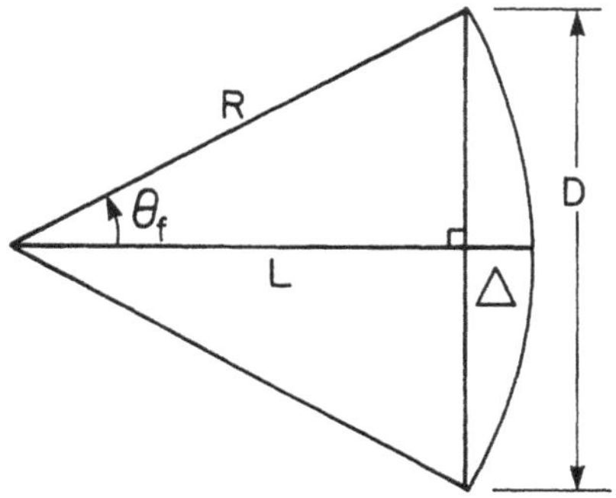

Figure A.12 An illustration of the sagitta, Δ.

Using the trigonometric identity

$$\tan\frac{\alpha}{2} = \frac{1 - \cos\alpha}{\sin\alpha}$$

we have

$$\Delta = R\sin\theta_f \tan\frac{\theta_f}{2} = \frac{D}{2}\tan\frac{\theta_f}{2} \tag{A.12}$$

We can develop an approximation that is valid in most cases. From Figure A.12,

$$R^2 = \left(\frac{D}{2}\right)^2 + (R - \Delta)^2 = \left(\frac{D}{2}\right)^2 + R^2 - 2R\Delta + \Delta^2$$

$$2R\Delta - \Delta^2 = \Delta(2R - \Delta) = \frac{D^2}{4}$$

If $\Delta \ll 2R$, which is true in almost all cases of interest,

$$\Delta \approx \frac{D^2}{8R} \tag{A.13}$$

This is a very useful approximation.

REFERENCES

1. D.G. Bodnar, "Materials and Design Data," Chapter 46 in *Antenna Engineering Handbook,* 2nd Ed. (R.C. Johnson and H. Jasik, eds.), New York: McGraw-Hill, 1984.
2. E.A. Mechtly, "Units, Constants, and Conversion Factors," Chapter 3 in *Reference Data for Engineers: Radio, Electronics, Computer, and Communications,* 7th Ed. (E.C. Jordan, ed.), Indianapolis, IN: Howard W. Sams, 1985.
3. R.V. Lowman, "Transmission Lines and Waveguides," Chapter 42 in *Antenna Engineering Handbook,* 2nd Ed. (R.C. Johnson and H. Jasik, eds.), New York: McGraw-Hill, 1984.
4. *ITE Antenna Handbook,* 2nd Ed., Philadelphia, PA: ITE Circuit Breaker Company, *circa* 1965.
5. T.S. Saad, R.C. Hansen, and G.J. Wheeler, *Microwave Engineer's Handbook,* Vol. 2, Norwood, MA: Artech House, 1971.
6. P.W. Hannan, "Microwave Antennas Derived from the Cassegrain Telescope," *IRE Trans. Antennas Propagat.,* Vol. AP-9, March 1961, pp. 140–153.
7. C.C. Cutler, "Parabolic-Antenna Design for Microwaves," *Proc. IRE,* Vol. 35, November 1947, pp. 1285–1295.
8. C.L. Gray, "Estimating the Effect of Feed Support Member Blocking on Antenna Gain and Side-Lobe Level," *Microwave J.,* March 1964, pp. 88–91.
9. K.S. Kelleher, "Reflector Antennas," Chapter 17 in *Antenna Engineering Handbook,* 2nd Ed. (R.C. Johnson and H. Jasik, eds.), New York: McGraw-Hill, 1984, pp. 17–18.
10. P.G. Hoel, *Introduction to Mathematical Statistics,* New York: John Wiley and Sons, 1947, p. 31.
11. R.B. Cotton, Georgia Tech Research Institute, Atlanta GA, private communication, 1982.
12. *American National Standard Safety Levels with Respect to Human Exposure to Radio Frequency Electromagnetic Fields, 300 kHz to 100 GHz,* ANSI C95.1-1982, New York: American National Standards Institute, 1982.

LIST OF SYMBOLS

A	frontal area of reflector
A_1, A_2	arbitrary amplitudes
C	Coulomb (electric charge)
D	diameter of reflector or length of a chord
D_b	diameter of small central blockage of a circular aperture
dB	decibel
dB_c	decibel level at coincident point in pattern comparisons
E	maximum field intensity of the unblocked aperture pattern
E_h	maximum field intensity of the hole pattern
E_s	maximum field intensity of the strut pattern
F	Farad (capacitance) or pattern field strength
f_c	cutoff frequency
ft	foot (length)
H	Henry (inductance)
J	Joule (energy)
K_g	constant for defining Gaussian pattern

K_s	constant for defining $(\sin x)/x$ pattern
L	length of a line-source aperture or length from the center of a circle to a cord
L_b	length of small central blockage of a line-source aperture
m	meter (distance)
mm	millimeter
mph	miles per hour
n, n	number
Np	neper
P_1, P_2	arbitrary powers
R	radius of circle
S	Siemens, electric conductance of a conductor in which a current of 1 A is produced by an electric potential difference of 1 V. (formerly called an "mho")
s	second (time)
V	wind velocity
W	width of aperture-blocking strut
Γ_r	field reflection coefficient
Γ_t	field transmission coefficient
Δ	sagitta length
$\Delta\phi$	phase shift
ϵ'/ϵ_0	dielectric constant
$\tan\delta$	loss tangent
θ_c	angle at coincident point in pattern comparisons
θ_f	half-flare angle
λ	wavelength
λ_c	cutoff wavelength

INDEX

The Artech House Microwave Library

Algorithms for Computer-Aided Design of Linear Microwave Circuits, Stanislaw Rosloniec

Analysis, Design, and Applications of Fin Lines, Bharathi Bhat and Shiban K. Koul

Analysis Methods for Electromagnetic Wave Problems, Eikishi Yamashita, ed.

Automated Smith Chart Software and User's Manual, Leonard M. Schwab

Capacitance, Inductance, and Crosstalk Analysis, Charles S. Walker

C/NL: Linear and Nonlinear Microwave Circuit Analysis and Optimization Software and User's Manual, Stephen A. Maas

Computer-Aided Design of Microwave Circuits, K.C. Gupta, R. Garg, and R. Chadha

Design of Impedance-Matching Networks for RF and Microwave Amplifiers, Pieter L.D. Abrie

Digital Microwave Receivers, James B. Tsui

Electric Filters, Martin Hasler and Jacques Neirynck

E-Plane Integrated Circuits, P. Bhartia and P. Pramanick, eds.

Evanescent Mode Microwave Components, George Craven and Richard Skedd

Feedback Maximization, Boris J. Lurie

Filters with Helical and Folded Helical Resonators, Peter Vizmuller

GaAs FET Principles and Technology, J.V. DiLorenzo and D.D. Khandelwal, eds.

GaAs MESFET Circuit Design, Robert A. Soares, ed.

GASMAP: Gallium Arsenide Model Analysis Program, J. Michael Golio, et al.

Handbook of Microwave Integrated Circuits, Reinmut K. Hoffmann

Handbook for the Mechanical Tolerancing of Waveguide Components, W.B.W. Alison

Handbook of Microwave Testing, Thomas S. Laverghetta

HEMTs and HBTs: Devices, Fabrication, and Circuits, Fazal Ali, Aditya Gupta, and Inder Bahl, eds.

High Power Microwave Sources, Victor Granatstein and Igor Alexeff, eds.

Introduction to Microwaves, Fred E. Gardiol

Introduction to Computer Methods for Microwave Circuit Analysis and Design, Janusz A. Dobrowolski

Introduction to the Uniform Geometrical Theory of Diffraction, D.A. McNamara, C.W.I. Pistorius, and J.A.G. Malherbe

LOSLIN: Lossy Line Calculation Software and User's Manual, Fred E. Gardiol

Lossy Transmission Lines, Fred E. Gardiol

MATCHNET: Microwave Matching Networks Synthesis, Stephen V. Sussman-Fort

Materials Handbook for Hybrid Microelectronics, J.A. King, ed.

Matrix Parameters for Multiconductor Transmission Lines: Software and User's Manual, A.R. Djordjevic, et al.

MIC and MMIC Amplifier and Oscillator Circuit Design, Allen Sweet

Microelectronic Reliability, Volume I: Reliability, Test, and Diagnostics, Edward B. Hakim, ed.

Microelectronic Reliability, Volume II: Integrity Assessment and Assurance, Emiliano Pollino, ed.

Microstrip Antenna Design, K.C. Gupta and A. Benalla, eds.

Microstrip Lines and Slotlines, K.C. Gupta, R. Garg, and I.J. Bahl

Microwave Circulator Design, Douglas K. Linkhart

Microwave Engineers' Handbook: 2 volume set, Theodore Saad, ed.

Microwave Filters, Impedance Matching Networks, and Coupling Structures, G.L. Matthaei, L. Young, and E.M.T. Jones

Microwave Integrated Circuits, Jeffrey Frey and Kul Bhasin, eds.

Microwaves Made Simple: Principles and Applications, Stephen W. Cheung, Frederick H. Levien, *et al.*

Microwave Materials and Fabrication Techniques, Second Edition, Thomas S. Laverghetta

Microwave MESFETs and HEMTs, J. Michael Golio, et al.

Microwave and Millimeter Wave Heterostructure Transistors and Applications, F. Ali, ed.

Microwave Mixers, Stephen A. Maas

Microwave Transmission Design Data, Theodore Moreno

Microwave Transition Design, Jamal S. Izadian and Shahin M. Izadian

Microwave Transmission Line Filters, J.A.G. Malherbe

Microwave Transmission Line Couplers, J.A.G. Malherbe

Microwave Tubes, A.S. Gilmour, Jr.

MMIC Design: GaAs FETs and HEMTs, Peter H. Ladbrooke

Modern GaAs Processing Techniques, Ralph Williams

Modern Microwave Measurements and Techniques, Thomas S. Laverghetta

Monolithic Microwave Integrated Circuits: Technology and Design, Ravender Goyal, *et al.*

Nonlinear Microwave Circuits, Stephen A. Maas

Optical Control of Microwave Devices, Rainee N. Simons

PLL: Linear Phase-Locked Loop Control System Analysis Software and User's Manual, Eric L. Unruh

Receiving Systems Design, Stephen J. Erst

Scattering Parameters of Microwave Networks with Multiconductor Transmission Lines: Software and User's Manual, A.R. Djordjevic, et al.

Solid-State Microwave Devices, Thomas S. Laverghetta

Stripline Circuit Design, Harlan Howe, Jr.

Terrestrial Digital Microwave Communications, Ferdo Ivanek, *et al.*

Time-Domain Response of Multiconductor Transmission Lines: Software and User's Manual, A.R. Djordjevic, et al.